Springer Geology

The book series Springer Geology comprises a broad portfolio of scientific books, aiming at researchers, students, and everyone interested in geology. The series includes peer-reviewed monographs, edited volumes, textbooks, and conference proceedings. It covers the entire research area of geology including, but not limited to, economic geology, mineral resources, historical geology, quantitative geology, structural geology, geomorphology, paleontology, and sedimentology.

More information about this series at http://www.springer.com/series/10172

Defan Guan · Xuhui Xu · Zhiming Li
Lunju Zheng · Caiping Tan
Yimin Yao

Theory and Practice of Hydrocarbon Generation within Space-Limited Source Rocks

石油工业出版社
PETROLEUM INDUSTRY PRESS

Defan Guan
Wuxi Research Institute of Petroleum
    Geology, RIPEP, SINOPEC
Wuxi
China

Xuhui Xu
Wuxi Research Institute of Petroleum
    Geology, RIPEP, SINOPEC
Wuxi
China

Zhiming Li
Wuxi Research Institute of Petroleum
    Geology, RIPEP, SINOPEC
Wuxi
China

Lunju Zheng
Wuxi Research Institute of Petroleum
    Geology, RIPEP, SINOPEC
Wuxi
China

Caiping Tan
Wuxi Research Institute of Petroleum
    Geology
Wuxi
China

Yimin Yao
Geological Scientific Research Institute
    of Shengli Oilfield Company, SINOPEC
Dongying
China

ISSN 2197-9545   ISSN 2197-9553   (electronic)
Springer Geology
ISBN 978-981-10-9608-2   ISBN 978-981-10-2407-8   (eBook)
DOI 10.1007/978-981-10-2407-8

Jointly published with Petroleum Industry Press, Beijing, China, 2016
Translation from the Chinese language edition: 烃源岩有限空间生烃理论与应用 by Defan Guan et al.,
© Petroleum Industry Press, Ltd., 2014. All Rights Reserved.

Preface

Hydrocarbon generation and accumulation within pore space limited source rocks are analyzed in this book. Our conclusions on characteristics of hydrocarbon generation and expulsion from pore space limited source rocks and hydrocarbon accumulations are based on the principles of basin formation, hydrocarbon generation, and accumulation in the practice of terrigenous basins in eastern China. Hydrocarbon generation and expulsion have been quantitatively evaluated in pore space limited source rock systems. Our new model of hydrocarbon generation and expulsion can provide a more accurate approach to the computation of petroleum yield within pore space limited source rocks and amount of expulsion. Results derived from our model are consistent with the subsurface geological situations. Theory and model based on our study have advantages over traditional kerogen thermal degradation and hydrocarbon generation models proposed by Tissot and Welte (1978).

This book will provide basic petroleum generation principles and practices, which can serve as new guide for all petroleum geologists, geochemists, teachers, and students in petroleum resource assessment.

Wuxi, China
2015

Defan Guan
Xuhui Xu
Zhiming Li

Contents

Introduction

Tissot and Welte published their pioneer work of *'Petroleum Formation and Occurrence'* in 1978. Chinese geochemists received this book as a gift from authors when they visited China in the same year. Their lectures and seminars were undoubtedly, to Chinese professionals, a timely summary of international scientific achievements. Once a Chinese edition of this book was published in 1982, the proposed theory of kerogen thermal degradation and hydrocarbon generation and analytical methods in the book were uncritically accepted by the Chinese petroleum geochemistry community, and then broadly applied to the practice of petroleum exploration, research and teaching in China. Until now, their theory and methods still prevail in Chinese organic geochemistry domain.

However, recent studies on Chinese national petroleum resource assessment gradually revealed that quantities of hydrocarbon generated and expelled from source rocks by the model based on this theory only reflect the maximum yield of oil and gas from a source rock at any time in experimental conditions rather than the actual amount of hydrocarbon generation at a certain maturity stage. The obvious flaw causes a mismatch between the calculated result and reality. Their theory is based on the assumption that all the organic matter was transformed into oil (or gas) by thermal degradation and no generation potential would remain at post mature stage, which is definitely disproved by the reality. Active carbon largely exists in present source rocks. Actually, the higher the quality of source rocks, the more the active carbon left in them. This suggests only a small proportion of kerogen was transformed during thermal degradation. Geochemists like Tissot are certainly aware of these facts. Then how did they justify the concept of source rock potential? They were inspired by oil shale pyrolysis generating shale oil. They regarded source rock thermal evolution the same as oil shale retorting and concluded that when temperature and time permitted, organic matter within source rocks would be transformed into hydrocarbon by thermal degradation as they were in oil shale pyrolysis.

In *'Petroleum Formation and Occurrence'*, Tissot and Welte (1978) regarded oil shale as an insoluble solid material (kerogen) and assessed its hydrocarbon generation potential using pyrolysis. They suggested that pyrolysis leading to shale oil formation from kerogen is comparable to the burial of the source rocks at depth where oil generation resulted from elevated temperature. No significant difference exists between hydrocarbon generation potential in natural evolution and the quantity of hydrocarbon generated by pyrolysis, because shallow, immature source rock (minimum organic richness required) shows no difference from oil shale.

Following the same theoretical consideration, Espitalié and other geochemists developed a source rock pyrolysis—an assessment device similar to oil shale destructive retorting, with temperature being the only controlling factor. This equipment has been widely applied to the experimental simulation of source rock thermal degradation, in which the maximum amount of hydrocarbon is generated (hydrocarbon generation potential is measured by $S_1 + S_2$) when temperature goes up to 550 °C. The S_1 measures volatile compounds in source rocks which are liberated at lower temperatures (below 300 °C), while S_2 measures remaining potential in insoluble kerogen, which does not degrade under a temperature below 550 °C. Based on Rock–Eval pyrolysis, a theory named kerogen thermal degradation was created.

What we can see from the above analysis is that this theory mainly deals with thermal degradation of organic matter (kerogen) by heating for hydrocarbon generation, which is undoubtedly flawless in terms of organic chemistry. However, hydrocarbon generation from source rocks should be an issue of petroleum geochemistry, which needs to be dealt with by means of petroleum geochemistry rather than organic chemistry. As is known to all, the principle of petroleum geochemistry is a combination of geology and chemistry (mainly organic chemistry) which seeks combined techniques from the perspective of geological conditions and settings to study relevant organic chemistry problems. Theory and principles of hydrocarbon generation from source rocks are typical problems of this category. Source rock developed in highly reducing and restricted depositional environments in sedimentary basins. The physicochemical transformation of organic matter during the geological history of sedimentary basins cannot be regarded as an isolated process. Hydrocarbon generation involves at least three parts: history of sediments evolution, thermal evolution of organic matter and physical–chemical interaction of oil, gas and water in a high temperature and high pressure regime. These parts are mutually associated and interacting throughout the whole process of source rock thermal evolution. Petroleum geochemistry principles rather than organic chemistry ones are required to study and analyze source rock degradation for hydrocarbon generation theory. Theories on hydrocarbon generation and its characters proposed in *Petroleum Formation and Occurrence* relied on pure experiments of oil shale pyrolysis with temperature and time as the only controlling variables, which take source rock away from natural subsurface conditions and environments, ignoring other environmental elements within source rock pores such as liquid pressure, porous water and clay minerals. Obviously, this theoretical consideration largely deviates from the aim of the discipline of petroleum geochemistry, resulting in the theory's deviation from reality and misleading in research and application.

To solve the problems in the current theory of kerogen thermal degradation for hydrocarbon generation, the authors and affiliated research groups spent more than a decade conducting petroleum geochemistry researches on hydrocarbon generation from pore space limited source rocks. Under the guidance of the theory of basin formation, hydrocarbon generation and accumulation, and by application of practical data from Shengli, Zhongyuan and Henan oilfields in eastern China, we independently developed a thermal-pressure controlled apparatus (type DK–I, DK–II and DK–III) to simulate hydrocarbon generation and expulsion under in situ natural evolution conditions. The preliminary results and applications will be summarized in the present book.

Chapter 1 is an introduction of petroleum geology theory. By an overview of petroleum geology history, analysis of traditional petroleum geological theories and petroleum system theories are conducted, and a systematic introduction of contents and research methods are involved.

Chapter 2 focuses on the current status of research on hydrocarbon generation from source rocks. Flaws of kerogen thermal degradation theory are discussed. Additionally, a comprehensive explanation of the principles of hydrocarbon generation and expulsion within pore space limited source rocks is presented.

Chapter 3 illustrates the characteristics of hydrocarbon generation and expulsion processes in pore space limited source rock and hydrocarbon accumulation. An introduction of erosion computation by means of Milankovitch cycles, the concept of sandstone relaxation and its applications are included.

Chapter 4 is the application of basin formation, hydrocarbon generation and accumulation in the Dongying Sag, eastern China.

Chapter 5 covers issues related to oil origins from marine and terrigenous depositional environments. Organic matter developed in marine source rocks from American and other major petroliferous basins are compared with those in terrigenous source rocks from China. Results show that both marine and terrigenous source rocks are capable of hydrocarbon generation and are inseparable parts of the theory of hydrocarbon generation. No essential difference can be illustrated in terms of hydrocarbon generation and accumulation.

Decades of research have been conducted by the present authors and affiliated research groups. We acknowledge the assistance from the Scientific Development Department of Sinopec Ltd. and considerable amounts of practical data from experts of Shengli, Zhongyuan and Henan oilfields. We are firmly determined that advances and innovations in petroleum geochemistry theories are the result of collaborative achievements made by petroleum geologists all over the world, and leading to the results of our work. For this reason, our foremost gratitude should be given to all petroleum geologists and petroleum geochemists.

Chapter 1
Brief Review of Petroleum Geology Theory Development

Petroleum geology theories are based on the practice of petroleum exploration and production and have been developed and improved by early geologists systematically sorting out first-hand data, introducing explanations and summarizing ideas. Therefore, they have been impressed with geological intelligence from the beginning and they comprise as highly a comprehensive discipline as other geological disciplines. Such comprehensiveness is characterized by multidisciplinary integration and cross inspirations among various subjects. By a brief overview of petroleum exploration history of the last 150 years, we can easily identify three major phases of its evolution.

1.1 Early Stage Petroleum Geology Theory

Petroleum is a mobile resource accumulated in the subsurface, which was first identified from surface oil seeps. Unsatisfied with the limited amount of oil seepage on the surface, venturers and explorers kept working on methods and techniques to find and obtain more petroleum from subsurface. At that time, oil seep formation and distribution principles caught the attention of geologists. They were convinced that oil seepage is leaking of subsurface oil accumulations up to the earth's surface through fractures and fissures under geological forces, suggesting that larger amounts of petroleum should be preserved in deep subsurface. They believed large amounts of petroleum could be recovered the same way as they got salt brine by drilling wells. In 1858, an American named E.L. Delark drilled the world's first oil well near an oil seep, found oil at 21.2 m depth and harvested a daily production of 4.1 tons. The success of this oil well inspired many others. Twenty-four oil wells were drilled in this region in just half a year. By analyzing large amounts of drilling data, geologists found that oil occurred in almost every well drilled on top of an anticline, while hardly at all in wells drilled on a syncline or horizontal strata. A geologist named T.S. Hunt formally proposed The Anticline Theory after two

© Petroleum Industry Press and Springer Science+Business Media Singapore 2017
D. Guan et al., *Theory and Practice of Hydrocarbon Generation within Space-Limited Source Rocks*, Springer Geology, DOI 10.1007/978-981-10-2407-8_1

years of research. An identical theory was proposed by another geologist named E.B. Anderlus in the same year. The main idea of this theory is the speculation of a subsurface container that is capable of preserving a petroleum accumulation. This sort of anticline is not the general type of but rather a vault, which has the shape of a upside down bowl. Petroleum is held in the upper portion by buoyancy as it is lighter than water. The theory was justified by the fact that petroleum tends to be found when drilling on the high position of anticline structures. The anticline theory was a hypothesis rather than a theory, which resulted from combined efforts of geologists by applying geology theories and analyzing numerous drilling data. Its academic value lies in a close involvement of the dominant issue of the concept of a trap in petroleum geology theories and a definite pinpoint of oil exploration target to anticlines. In the early phase of oil exploration, the US Geology Survey conducted a nationwide geological survey searching for anticlines, applying this exploration method. Their endeavor not only effectively improved the success rate of oil exploration, but also pushed forward exploration, resulting in the USA's annual petroleum production reaching 100,000,000 tons.

Although the anticline theory was a hypothesis at the early phase of petroleum exploration, it was a guide in petroleum exploration practices for the following hundred years, with a transformation of anticline discovery techniques from field geological surveying to seismic profiling. Currently, global petroleum exploration follows the regular routine of 'seismic techniques for traps, drilling for petroleum and geology for understanding'. Up until now, about 80 % of oil fields and over 90 % of gas fields have been discovered in anticlinal traps. Practice testified the theory, which was established as the typical petroleum geological theory in the early phase.

As petroleum production increased with time, geologists were increasingly recruited by petroleum enterprises to become involved in petroleum exploration. Inspired by this theory, many geologists set down to working on petroleum geology theories, among which theories of anticlinal traps, marine origin of petroleum, porosity and permeability of reservoir rocks and fixed carbon ratio successively resulted as the outcome of their endeavors. In 1917, the American Association of Petroleum Geology (AAPG) was established. Along with it was the emergence and prevalence of relevant disciplines and exploration technologies, such as the disciplines of biostratigraphy, micropaleontology, petrology, well logging, seismic surveying and well testing, which were put into broad application. By the 1930s, contemporary concepts of petroleum geology had come into being. Ironically, many geologists refused to accept petroleum geology as an academic discipline despite their success in petroleum exploration. They regarded petroleum exploration as merely practices and continuous refinements of geological theories by virtue of various methods (such as drilling, logging and well testing). Geology belongs to the category of science while petroleum exploration is a skilled work which relies on technology. Every geologist once systematically trained can be involved in petroleum exploration. No petroleum geologist existed at early time but geologists and petroleum engineers did. This situation not only inhibited overall development of petroleum geology theories, strangling it in the bud, but also oriented the consequent hundred years of petroleum exploration to research on geology and relevant disciplines. The trend continues even until today, preventing petroleum geology from all-around developments.

1.2 Origin and Development of Traditional Petroleum Geology Theories

In 1954, American geologist A.I. Levorsen had his treatise *Geology of Petroleum* published. His book elaborated issues of petroleum's organic origin, migration, accumulation and trap types, from the perspectives of traditional geological ideas, based on analyses of extensive data of one hundred years of global petroleum exploration. The most remarkable idea of his book was the identification of the seven elements of 'generation, reservoir, seal, migration, accumulation, trap and preservation' as the fundamental elements for hydrocarbon accumulations coming into being in sedimentary basins. These ideas not only act as effective guidance for petroleum exploration in sedimentary basins, but also inspired geologists' theoretical thinking patterns to study petroleum exploration. For years, his proposed way of thinking which has been adopted by petroleum explorers worldwide to study sedimentary basins and evaluate their prospects. Thus, A.I. Levorsen was regarded as 'the father of petroleum geology theory'.

In 1973 and 1983, Dr. R.E. Charpton from Queensland University, Australia published the first and second edition of his book *Petroleum Geology*. The books featured a discussion of petroleum generation, migration and accumulation mechanics in terms of geosciences and physics, and an unprecedented summary of the regular patterns of the age and geotectonic distribution of oil and gas, by means of sedimentary basin evolution, structural deformation and stratigraphic variations. His ideas and knowledge not only enriched the contents of Levorsen's book, but also provided good guidance to petroleum geology research.

In 1985, based on over three decades of practical experience in teaching and working in petroleum enterprises, Canadian Professor F.K. North spent three years on his work of *Petroleum Geology*, which has been considered as the world's most comprehensive and systematical text book. In his work of 910,000 words, North made a comprehensive summary of previous research achievements on petroleum properties and origins, accumulation, occurrence and exploration techniques, as well as systematically illustrating his theories. In addition to that, he also learned a detailed lesson from contemporary oil and gas production practice. This book is a good summary of traditional theories on petroleum geology as well as lessons learnt from previous practice of petroleum exploration and prospecting.

These three geological treatises published during the 1950s and 1980s generally embrace the essential ideas of traditional theories, and share the common feature of addressing the issues of petroleum characterization and origin analysis by basic geological principles and geological wisdom. The main ideas are practical experiences of petroleum exploration, coupled with multidiscipline rules in areas like geochemistry, hydrodynamics and thermo-kinetics, etc. Thus, traditional geological theories are a combination of geology, petroleum geology and other relevant disciplines, rather than full-scale methods that can be applied to petroleum characterization. In application of the theories, the obvious defects appear as briefly illustrated here:

① Petroleum geology theory as whole was undermined by numerous subdivisions which are relevant but not integrated, leading to an all-sided development but failure to be a comprehensive petroleum geology theory.

The core of traditional petroleum geology theories is the analysis and assessment of potential of hydrocarbon accumulations, namely the seven elements: generation, reservoir, seal, migration, accumulation, trapping and preservation. This concept proves infallible not only theoretically but also practically. However, the knowledge and techniques in assessment remains a big challenge for most petroleum explorers in a practical way as they involve multidisciplinary theories and methods, which are beyond the scope that a petroleum geologist can handle. The seven domains of scientific work are separately assigned to specialized fields. The obvious consequence is everyone works for his own part without integrity. Those who have expertise in petroleum generation mainly work on source rock occurrence in sedimentary basins, petrology, geochronology and petroleum geochemistry, and gradually become petroleum geochemists. People in reservoir and seal research focus their work on sedimentology and sequence stratigraphy and become experts in this field. Those who have expertise in migration and accumulation work on oil–gas–water dynamics and potential energy field in a sedimentary basin and attempt to be an expert in this field. Trap and preservation researchers mainly work on structural geology and tectonic characterization of sedimentary basins and trap formation mechanisms. The comprehensive analysis of sedimentary basin prospecting is split into seven isolated parts that is dealt with by relevant professionals. This practice in the application of traditional petroleum geology theories, in effect, is a 'disassembly' of petroleum geology theory to make it a jigsaw puzzle consisting of various disciplines. This terrible situation has been taking its toll until even today the traditional theories are less likely to be advanced.

② Traditional petroleum geological theory relies on geological analytical methods and data generated from all measurements to describe geological phenomena statically, which does not match the dynamic process of hydrocarbon accumulations.

The seven essential elements of the formation of hydrocarbon accumulations are the most fundamental idea of traditional petroleum theories, which is an all-embracing, regulatory summary of production practices in discovered reservoirs. However, the availability of these seven elements does not necessarily lead to hydrocarbon accumulations in the subsurface. The optimal combination of these elements is essential. Among the seven key elements, source rock is undoubtedly the most important one. Consequently, the priority is to determine the generation and expulsion timing of the source rock, based on dynamic analysis of reservoir conditions and possible migration pathways. In the light of that knowledge, assessment of trap and preservation conditions and the consequent analysis of pros and cons of hydrocarbon accumulations can be done in great detail and finally give clues to the prediction of exploration targets. The character of mobility makes petroleum resources distinct from solid resources, which render hydrocarbon generation and expulsion from source rock the core, and dynamic analysis of the seven elements' correspondence the essential content of the research on hydrocarbon accumulations.

1.3   The Petroleum System and Its Characteristics

The term petroleum system was formally proposed by American petroleum geochemist L.B. Magoon et al. in 1987. Its main characteristic is an integration and redefinition of the seven elements into two categories, i.e., elements and processes. Four elements are source rock, reservoir, seal and overburden. The processes are dynamically encompassing hydrocarbon generation, migration, accumulation and preservation. In brief, a petroleum system encompasses a pod of active source rock and all genetically related oil and gas accumulations. It includes all the geologic elements and processes that are essential if an oil and gas accumulation is to exist. The term "petroleum" here describes a compound that includes high concentration of any of the following substances: thermal and biological hydrocarbon gas found in conventional reservoirs as well as in gas hydrates, tight reservoirs, fractured shale and coal, condensates, crude oils and natural bitumen in reservoirs, generally in siliciclastic and carbonate rocks. The term "system" describes the interdependent elements and processes that form the functional unit that creates hydrocarbon accumulations. These essential elements and processes must be correctly placed temporally and spatially so that organic matter included in a source rock can be converted into a petroleum accumulation. For an illustration of effective source rock-petroleum accumulation geochemical correlations, a distinct analysis of petroleum system geography, stratigraphy and temporal and spatial extent of a petroleum system and a final determination of generation-accumulation efficiency, the investigation of petroleum system also includes four figures (burial history chart, petroleum system map, petroleum system cross section, events chart) and one table, a table of hydrocarbon accumulations.

The so-called "critical event" refers to the most properly chosen moment when major generation migration accumulation process of a petroleum system takes place. The petroleum system map and cross section of critical moment can best show the geographical and stratigraphic extent of the system. The critical moment when the majority of hydrocarbons is generated, migrating and accumulating in the primary traps can be induced via analysis of a burial history figure.

The research on hydrocarbon accumulations by means of a petroleum system can be illustrated by charts and table of petroleum system events. In other words, it means to readdress the seven elements in traditional petroleum theories, i.e., generation, reservoir, seal, migration, accumulation, trap and preservation as petroleum system events. Source rock, reservoir, seal and overburden are fundamental elements formed during deposition, while trap formation and hydrocarbon generation, migration, accumulation and preservation are dynamic processes which reveal the emplacement sequence taking place within a petroleum system. Geophysical data and tectonics analysis is applied to search for traps. Research on time required for the process of generation migration accumulation of hydrocarbons, namely the age of a petroleum system, are based on the research of stratigraphy, petroleum geochemistry and burial history. Preservation time begins immediately after the generation–migration–accumulation process occurs and extends to the present day. It encompasses any risks to the petroleum accumulations during this period. An incomplete or just completed petroleum system lacks a preservation time. It is valid to consider that a large proportion of hydrocarbon within the petroleum system can be preserved until now. The critical moment is determined via burial history, the same as timing in petroleum map and cross section.

The concept and describing methods of a petroleum system not only put the complicated practical data into the simple form of four figures and one table, but also provide an easy way of interpretation and application for petroleum geologists. More importantly, it puts hydrocarbon generation and accumulation as inter-connected components to study the prerequisites in an integrated manner which combine dynamics and statics, description and measurement, time and space. Hence, the concept of petroleum system has received great acknowledgements ever since the proposal of it. Judging from the single aspect of the application of four graphs and one table, it is merely regarded as a research method by petroleum geologists. But from a theoretical thought perspective, it brings up a couple of thought provoking issues, which is more valuable. '*Petroleum System*' has a lot of descriptions, which put basic elements of petroleum system into strata distribution in the region. The source rock at the critical moment is linked to reservoir, seal and overburden rock. The overburden rocks, not included in traditional petroleum geology, not only serve as prerequisite for source rock thermal maturation, but also shape the geometry of underlying conduits for hydrocarbon migration and trapping. Although many questions remain to be answered regarding the involvement of overburden, it is geological integrity for a petroleum accumulation. Another example is the description of basin dynamics. Source rocks play an important role in petroleum generation, trapping and preservation. However, the formation of a petroleum system results from interactions of various physical and chemical factors (diagenesis, structural deformation and compaction). These processes have various effects on sediments and take direct controls on hydrocarbon generation, accumulation and destruction. The controlling factor is structural adjustment of sedimentary basins. We further emphasize the triggering factors are tectonics, heat flux and gravity. The formation of a petroleum system is the interaction of various physical and chemical factors and structural adjustment is the critical factor controlling hydrocarbon accumulations. The authors of *Petroleum System* admitted that some of their concepts are hard to understand and further investigations are required. We admire their modest attitude and explorative spirits as they are asking insightful questions concerning the connotations of hydrocarbon generation and hydrocarbon accumulations, which add significant academic values in theory of petroleum system.

The theory of petroleum system is the first attempt to address the issue of a petroleum system in an integrated manner to consider source rock and hydrocarbon accumulation as a whole. As L.B. Magoon has stressed, the focus of petroleum system research is the illustrations of causal relationships between mature source rock and correlated hydrocarbon accumulations, naming from source rock to traps. However, the theory of petroleum system fails to further probe into the connection of hydrocarbon generation and accumulation with sedimentary basin development and evolutional characteristics. In spite of the awareness of direct causal relationships between tectonic adjustments and petroleum system, no breakthrough in commitment of triggering events and tectonic adjustments controlling hydrocarbon accumulations has ever been documented.

1.4   Principles of Basin Formation, Hydrocarbon Generation and Accumulation

According to the point of petroleum's organic origin, petroleum geologists generally regard the sedimentary basin as the basic unit for petroleum accumulation as long as hydrocarbon generation, migration and accumulation have taken place within its domain, regardless of differences in basin size, morphological characteristics and formation mechanisms. Therefore, all petroliferous basins share the same or similar characteristics of petroleum geological evolution and development. By serious analysis of petroleum geological characteristics of sedimentary basins at different evolution stages, coupled with inter-relationships of hydrocarbon generation and accumulation, the process of basin formation, hydrocarbon generation and accumulation can be integrated as a whole in the discussion of various geological problems.

The geological evolution history of 98 Mesozoic–Cenozoic petroliferous basins (sags or depressions) (Table 1.1) indicate that all those basins have gone through three distinct phases: subsidence, uplift and shrinkage, despite of differences in their origin, morphological characteristics and sizes.

1.4.1   Basin Subsidence Stage

Patterns of a sedimentary basin in the subsidence stage are characterized by either depression or rift, with a brief uplifting or even erosion during whole process. The general characteristics of petroliferous basins in this phase are:

① High rate of subsidence

The subsidence rate of most sedimentary basins is more than 200 m/Ma with most important source rock development in the basin. For example, the subsidence rate

Table 1.1 Development characteristics of 98 Mesozoic–Cenozoic petroliferous basins (depressions or sags) worldwide

Basin (or depression) name	Basin development stage	Subsidence	Uplifting	Shrinkage
1. Songliao Basin	Cretaceous–Quaternary	Quantou–Nenjiang Fm deposition stage	End of Nenjiang Fm deposition	Sifangtai Fm deposition stage–Quaternary
2. Hailaer Basin	Cretaceous–Quaternary	Tongtimiao Fm–Yimin Fm deposition stage	End of Yimin Fm deposition	Qingyuangang Fm deposition stage–Quaternary
3. Kailu Depression	Jurassic–Quaternary	Late Jurassic	End of Fuxin Fm deposition	Yaojia Fm deposition stage–Quaternary
4. Yilan–Yitong Basin	Paleogene–Quaternary	Paleogene	End of Qijia Fm deposition	Chaluhe Fm deposition stage–Quaternary
5. Erlian Basin	Cretaceous–Quaternary	Early Cretaceous	End of Saihantala Fm deposition	Late Cretaceous–Quaternary
6. Liaohe Depression	Paleogene–Quaternary	Paleogene	End of Dongying Fm deposition	Guantao Fm deposition stage–Quaternary
7. Huangye Depression	Paleogene–Quaternary	Paleogene	End of Dongying Fm deposition	Guantao Fm deposition stage–Quaternary
8. Jizhong Depression	Paleogene–Quaternary	Paleogene	End of Dongying Fm deposition	Guantao Fm deposition stage–Quaternary
9. Jiyang Depression	Paleogene–Quaternary	Paleogene	End of Dongying Fm deposition	Guantao Fm deposition stage–Quaternary
10. Dongpu Depression	Paleogene–Quaternary	Paleogene	End of Dongying Fm deposition	Guantao Fm deposition stage–Quaternary
11. Miyang Depression	Paleogene–Quaternary	Paleogene	End of Liaozhuang Fm deposition	Shangsi Fm deposition stage–Quaternary
12. Subei Basin	Paleogene–Quaternary	Paleogene	End of Sanduo Fm deposition	Yancheng Fm deposition stage–Quaternary
13. Jianghan Basin	Paleogene–Quaternary	Paleogene	End of Jinghezhen Fm deposition	Guanghuasi Fm deposition stage–Quaternary
14. Sanshui Basin	Paleogene–Quaternary	Paleogene	End of Huayong Fm deposition	Eocene–Quaternary
15. Baise Basin	Paleogene–Quaternary	Paleogene	End of Jianduling Fm deposition	Oligocene–Quaternary

(continued)

Table 1.1 (continued)

Basin (or depression) name	Basin development stage	Subsidence	Uplifting	Shrinkage
16. Fushan Depression	Paleogene–Quaternary	Paleogene	End of Weizhou Fm deposition	Neogene–Quaternary
17. Zhujiangkou Basin	Paleogene–Quaternary	Paleogene	End of Zhuhai Fm deposition	Neogene–Quaternary
18. Qiongdongnan Basin	Paleogene–Quaternary	Paleogene	End of Lingshui Fm deposition	Neogene–Quaternary
19. Yinggehai Basin	Paleogene–Quaternary	Paleogene	End of Lingshui Fm deposition	Neogene–Quaternary
20. Beibuwan Basin	Paleogene–Quaternary	Paleogene	End of Weizhou Fm deposition	Neogene–Quaternary
21. Jiuquan Basin	Cretaceous–Quaternary	Cretaceous	End of Cretaceous	Oligocene–Quaternary
22. Tuha Basin	Jurassic–Quaternary	Jurassic	End of Jurassic	Cretaceous–Quaternary
23. Ordos Basin	Late Triassic–Early Cretaceous	Late Triassic	End of Late Triassic	Jurassic–Early Cretaceous
24. Junggar Basin	Late Paleozoic–Quaternary	Late Permian–Jurassic	End of Jurassic	Cretaceous–Quaternary
25. Qaidam Basin	Paleogene–Quaternary	Paleogene–Neogene	End of Neogene	Quaternary
26. Yanqi Basin	Triassic–Quaternary	Late Triassic–Jurassic	End of Jurassic	Cretaceous–Quaternary
27. Sichuan Basin	Late Triassic–Quaternary	Late Triassic–Early Jurassic	End of Middle Jurassic	Late Jurassic–Quaternary
28. East China Sea Basin	Paleogene–Quaternary	Paleogene	End of Huagang Fm deposition	Neogene–Quaternary
29. Santanghu Basin	Late Paleozoic–Quaternary	Late Permian–Middle Jurassic	End of Xishanyao Fm deposition	Middle Jurassic–Quaternary
30. Baiyunchagan Depression	Cretaceous–Quaternary	Early Cretaceous	End of Saihantala Fm deposition	Late Cretaceous–Quaternary
31. Jinggu Basin	Paleogene–Quaternary	Paleogene	End of Mengla Fm deposition	Neogene–Quaternary
32. Lunbola Basin	Paleogene–Quaternary	Paleogene	End of Dingqinghu Fm deposition	Neogene–Quaternary
33. Jidong Depression	Paleogene–Quaternary	Paleogene	End of Dongying Fm deposition	Guantao Fm deposition stage–Quaternary
34. Kuche Depression	Triassic–Quaternary	Middle Triassic–Jurassic	End of Late Jurassic deposition	Cretaceous–Quaternary

(continued)

Table 1.1 (continued)

Basin (or depression) name	Basin development stage	Subsidence	Uplifting	Shrinkage
35. New Zealand west coast Basin	Cretaceous–Quaternary	Late Cretaceous–Oligocene	End of Oligocene	Neogene–Quaternary
36. Australian Gippsland Basin	Late Cretaceous–Quaternary	Late Cretaceous–Eocene	End of Eocene	Oligocene–Quaternary
37. Australian Burroughs Basin	Late Cretaceous–Quaternary	Late Cretaceous–Oligocene	End of Oligocene	Neogene–Quaternary
38. Australian Watervi Basin	Jurassic–Quaternary	Late Jurassic–Cretaceous	End of Late Cretaceous	Paleogene–Quaternary
39. Luowan–Maleic Basin	Eocene–Quaternary	Eocene–Miocene	End of Miocene	Pliocene–Quaternary
40. Wenlai–Shaba Basin	Eocene–Quaternary	Eocene–Miocene	End of Miocene	Pliocene–Quaternary
41. Middle Sumatra Basin	Eocene–Quaternary	Eocene–Oligocene	End of Oligocene	Miocene–Quaternary
42. Kouté Basin (Indonesia)	Eocene–Quaternary	Eocene–Oligocene	End of Oligocene	Miocene–Quaternary
43. Mekong Delta Basin	Eocene–Quaternary	Eocene–Miocene	End of Miocene	Pliocene–Quaternary
44. Potwar Basin	Paleogene–Quaternary	Paleocene–Eocene	End of Eocene	Pliocene–Quaternary
45. Indian River Basin	Cretaceous–Quaternary	Cretaceous–Miocene	End of Miocene	Pliocene–Quaternary
46. Bombay Basin	Paleogene–Quaternary	Paleocene–Miocene	End of Miocene	Pliocene–Quaternary
47. Man Ghosh Lark Basin (Middle Asia)	Triassic–Quaternary	Triassic–Early Cretaceous	End of Early Cretaceous	Late Cretaceous–Quaternary
48. South Caspian Sea Basin	Jurassic–Quaternary	Jurassic–Miocene	End of Pliocene	Quaternary
49. Karakum Basin	Jurassic–Quaternary	Jurassic–Cretaceous	End of Cretaceous	Paleocene–Quaternary
50. Turgay Basin	Jurassic–Quaternary	Early Jurassic		Cretaceous–Quaternary
51. Fergana Basin	Jurassic–Quaternary	Late Cretaceous–Paleocene	End of Oligocene	Neogene–Quaternary
52. Kuhla Basin	Cretaceous–Quaternary	Cretaceous–Oligocene	End of Oligocene	Pliocene–Quaternary
53. Yemen Qamar–Jeza Basin	Cretaceous–Quaternary	Middle–Late Cretaceous	End of Late Cretaceous	Paleogene–Quaternary
54. Yemen Ma'rib–Shabwah Basin	Late Jurassic–Quaternary	Late Jurassic–Late Cretaceous	End of Late Cretaceous	Paleogene–Quaternary
55. Brazil Toguar Basin	Late Jurassic–Quaternary	Late Jurassic–Late Cretaceous	End of Late Cretaceous	Paleogene–Quaternary
56. North Sea Basin (England)	Late Jurassic–Quaternary	Late Jurassic–Late Cretaceous	End of Late Cretaceous	Paleogene–Quaternary

(continued)

Table 1.1 (continued)

Basin (or depression) name	Basin development stage	Subsidence	Uplifting	Shrinkage
57. Benue Basin	Cretaceous–Quaternary	Cretaceous–Paleogene	End of Paleogene	Neogene–Quaternary
58. Ogaden Basin	Jurassic–Quaternary	Early Jurassic–Cretaceous	End of Late Cretaceous–Oligocene	Late Oligocene–Quaternary
59. Suez Basin	Late Oligocene–Quaternary	Late Oligocene–Early Miocene	End of Early Miocene	Pliocene–Quaternary
60. The Nile River Basin	Mesozoic–Quaternary	Late Cretaceous–Early Miocene	End of Early Miocene	Pliocene–Quateranry
61. Siirt Basin	Cretaceous–Quaternary	Middle Paleocene	End of Middle Paleocene	Late Eocene–Quaternary
62. Sheriff Basin (Algeria)	Miocene–Quaternary	Miocene–Pliocene	End of Pliocene	End of Pliocene–Quaternary
63. Cuanza Basin	Late Jurassic–Quaternary	Late Jurassic–Miocene	End of Miocene	Pliocene–Quaternary
64. Muglad Basin	Early Cretaceous–Quaternary	Early–Late Cretaceous	Late Cretaceous–Paleocene	Neocene–Quaternary
65. Mozambique Basin	Jurassic–Quaternary	Late Jurassic–Cretaceous	End of Late Jurassic	Paleocene–Quaternary
66. MulunDawa Basin	Middle Jurassic–Quaternary	Late Jurassic–Late Cretaceous	End of Late Cretaceous	Miocene–Quaternary
67. Abidjan Basin	Jurassic–Quaternary	Jurassic–Cretaceous	End of Cretaceous	Paleocene–Quaternary
68. Niger Delta Basin	Late Cretaceous–Quaternary	Later Cretaceous–Oligocene	End of Oligocene	Neocene–Quaternary
69. Low Congo Basin	Late Cretaceous–Quaternary	Late Cretaceous–Oligocene	End of Oligocene	Neocene–Quaternary
70. Dopa Basin	Cretaceous–Quaternary	Cretaceous	End of Cretaceous	Paleocene–Quaternary
71. Mozambique Basin	Jurassic–Quaternary	Jurassic–Early Cretaceous	End of Late Cretaceous	Paleocene–Quaternary
72. Marenjia Basin	Middle Jurassic–Quaternary	Middle Jurassic–Late Cretaceous	End of Late Cretaceous	Miocene–Quaternary
73. Orange Basin	Late Cretaceous–Quaternary	Late Jurassic–Cretaceous	End of Cretaceous	Paleogene–Quaternary
74. Hermosi Basin	Early Jurassic–Quaternary	Early Jurassic–Oligocene	End of Oligocene	Miocene–Quaternary
75. Pannonia Basin	Middle Eocene–Quaternary	Middle Eocene–Early Miocene	End of Early Miocene	Miocene–Quaternary
76. Carpathian Basin	Jurassic–Quaternary	Late Jurassic–Miocene	End of Miocene	Pliocene–Quaternary
77. Upper Rhein Graben Basin	Middle Eocene–Quaternary	Middle Eocene–Early Miocene	End of Early Miocene	End of Miocene–Quaternary

(continued)

Table 1.1 (continued)

Basin (or depression) name	Basin development stage	Subsidence	Uplifting	Shrinkage
78. Former Apennines Basin	Jurassic–Quaternary	Jurassic–Miocene	End of Miocene	Pliocene–Quaternary
79. Paris Basin	Jurassic–Quaternary	Jurassic–Eocene	End of Eocene	Miocene–Quaternary
80. Zapadnays Sibir' Basin	Jurassic–Quaternary	Jurassic–Early Cretaceous	End of Cretaceous	Miocene–Quaternary
81. North Ciscaucasia Basin	Cretaceous–Quaternary	Cretaceous–Early Miocene	End of Early Miocene	Late Miocene–Quaternary
82. Sahalim Basin	Paleogene–Quaternary	Paleocene–Miocene	End of Miocene	Pliocene–Quaternary
83. Albert Basin	Jurassic–Quaternary	Jurassic–Paleocene	End of Paleocene	Pliocene–Quaternary
84. Los Angeles Basin	Neogene–Quaternary	Neogene–Early Pliocene	End of Early Pliocene	Late Pliocene–Quaternary
85. Mexico Gulf Basin	Late Jurassic–Quaternary	Late Jurassic–Oligocene	Late Oligocene	Miocene–Quaternary
86. Tampico Basin	Jurassic–Quaternary	Jurassic–Cretaceous	End of Cretaceous–Prime Paleocene	Paleocene–Quaternary
87. North Basin	Jurassic–Quaternary	Jurassic–Cretaceous	End of Cretaceous–Prime Paleocene	Paleocene–Quaternary
88. San Juan Basin	Middle Jurassic–Quaternary	Middle Jurassic–Cretaceous	End of Cretaceous	Paleogene–Quaternary
89. Ewing tower Basin	Middle Jurassic–Quaternary	Middle Jurassic–Cretaceous	End of Cretaceous	Paleogene–Quaternary
90. Acturus Basin	Middle Jurassic–Quaternary	Middle Jurassic–Cretaceous	End of Cretaceous	Paleogene–Quaternary
91. Arctic Slope Basin	Permian–Quaternary	Permian–Early Cretaceous	End of Early Cretaceous	Late Cretaceous–Quaternary
92. San Joaquin Basin	Eocene–Quaternary	Eocene–Miocene	End of Miocene	Pliocene–Quaternary
93. Ventura Basin	Cretaceous–Quaternary	Cretaceous–Pliocene	End of Pliocene	Pleistocene–Quaternary
94. Powder River Basin	Cretaceous–Quaternary	Cretaceous–Pliocene	End of Pliocene	Pleistocene–Quaternary
95. Glenn Duttan Basin	Late Triassic–Quaternary	Late Triassic–Early Cretaceous	End of Early Cretaceous	Late Cretaceous–Quaternary
96. Maracaibo Basin	Late Jurassic–Quaternary	Cretaceous–Oligocene	End of Oligocene	Miocene–Quaternay
97. Putumayo Basin	Jurassic–Quaternary	Jurassic–Cretaceous	End of Cretaceous	Paleogene–Quaternary
98. Magellan Basin	Late Jurassic–Quaternary	Late Jurassic–Paleocene	End of Paleocene	Eocene–Quaternary

of the second and third member of the Hetaoyuan Formation in the Miyang Depression is as high as 394 m/Ma; the subsidence rate of the upper forth and the third member of the Shahejie Formation in the Dongying Sag is as high as 300 m/Ma; subsidence rate of Monterey Formation and Repetto Formation, the major source rock in the Los Angeles Basin is as high as 430 m/Ma.

② High TOC content forming the major source rock intervals

Residual TOC contents of source rocks in this stage of basin development are largely above 1 % with the most organic-rich intervals greater than 2 %. For example, source rocks in the first member of the Qingshankou Formation in the Songliao Basin have an average residual TOC content of 2.24 %. Source rocks of Monterey Formation and Repetto Formation in Los Angeles Basin have an average residual TOC content of 3.12 %. The Sirte Shale of the Libyan Sirte Basin has a residual TOC content over 5 %.

③ Most source rocks reached hydrocarbon generation threshold and occasionally hydrocarbon generation process was completed during the last phase of subsidence

Due to relatively high subsidence and deposition rates during the subsidence stage, sediments accumulated in the sedimentary basins can be several thousand meters thick. The temperature gradients of these basins are mostly well beyond the normal level, so in the last phase of basin subsidence, major source rocks have largely reached hydrocarbon generation threshold and matured. In other words, during this stage of petroliferous basins, major source rocks have gone through the thermal evolution under high temperature–pressure regime and completed hydrocarbon generation (Table 1.2).

Table 1.2 Maturity of main source rocks by the end subsidence phase

Basin	Age at the end of subsidence	Burial depth of the main source rock base	Oil window interval	Geothermal gradient (°C/km)
Songliao Basin	Late deposition of Nenjiang Fm	Burial depth of the 1st member of Qingshakou Fm is 2000 m	1200–2100 m, Ro 0.5–1.2 %	42
Bohai Bay Basin (Jiyang depression)	Late deposition of Dongying Fm	Burial depth of the upper ES4 member is 3000–3500 m	2200–3800 m, Ro 0.5–1.2 %	36
Nanxiang Basin (Biyang depression)	Late deposition of Liaozhuang Fm	Burial depth of the 3rd member of Hetaoyuan Fm is 3500 m	1800–3100 m, Ro 0.5–1.2 %	41
Jianghan Basin	Late deposition of Jinghezheng Fm	Burial depth of the 1st member of Xinjiagou and the 4th and 3rd members of Qianjiang are 4694, 3441 and 2096 m.	2500–3700 m, Ro 0.5–1.2 %	34
Los Angeles Basin	Late deposition of Leipeituo Fm	Burial depth of the Mengtelei Fm is 4800 m	2500 m, Ro about 1 %	39.1

1.4.2 Basin Uplift Stage

After the subsidence stage, sedimentary basins will go through a stage of uplifting and erosion. While multiple factors may cause basin reversal development, geologists' main task is to probe into characteristics of this phase and their effects on hydrocarbon generation and accumulation. An integrated analysis could get us to the summary of basin's basic characteristics in this stage:

① Different degree of erosion of sedimentary strata

The erosion process in the uplift stage lasts millions of years, the eroded strata thickness is in the range of hundreds to thousands of meters and erosion rate is tens to hundreds of m/Ma. For example, erosion lasted for 10.6 Ma in the uplift stage of the Dongying Sag during the late period of its development, with the eroded strata thickness being around 1000 m and erosion rate being around 94 m/Ma. The erosion time lasted for 18 Ma in the Gaoyou Sag of the Subei Basin during the uplift stage of basin development, with the eroded strata thickness of 1500–2000 m and average erosion rate around 97 m/Ma.

② Various erosion rates and eroded amounts at different structures

During the uplift stage of basin development, erosion rate and eroded thickness of strata differ greatly at different structural locations. In general, erosion rates are higher and eroded amounts are larger in the basin margin sandstone rich region; while erosion rates are lower and eroded amounts are smaller in the mudstone rich zone. Still some area such as depocenter may be characterized by slow subsidence rather than uplift during this stage. For instance, in the uplift stage when sedimentation is almost ended, the fourth and fifth members of the Nenjiang Formation in the marginal area of the Songliao Basin have been eroded completely, while only the fifth member of the Nenjiang Formation has been eroded in the Changyuan region and almost no erosion occurs in the Gulong Sag.

③ A crucial episode and an indicator of hydrocarbon accumulation in the basin

Based on data analyses of 98 Mesozoic–Cenozoic petroliferous basins all over the world, the uplifting stage of basin development is a crucial episode of hydrocarbon accumulation and this period is an indicative of hydrocarbon accumulation in the basin. For instance, uplifting and erosion of the Nenjiang Formation in the Songliao Basin is the main phase for hydrocarbon accumulation to form the giant Daqing Oilfield. Uplifting and erosion of the Dongying Formation in the Bohai Bay Basin is the main phase for hydrocarbon accumulation in all sags (depressions). Uplifting and erosion process of the Pliocene Repetto Formation in the Los Angeles Basin is the main phase for hydrocarbon accumulation in the basin's largest oilfield, Wilmington Oilfield. Uplifting and erosion process of the Miocene Trisha Formation in the Indonesia Central Sumatra Basin is the main phase for hydrocarbon accumulation in the basin's largest oilfield—Minas Oilfield. In summary, petroliferous basins have all gone through millions of years of uplift and erosion, accompanied by hydrocarbon generation, migration and accumulation, which symbolizes the hydrocarbon accumulation stage in petroliferous basins.

1.4.3 Basin Shrinkage Stage

After the subsidence and uplift stages, petroliferous basins experience a period of overall shrinkage, which normally lasts till the Quaternary.

① Small scale of uplift and subsidence

Basin development during this time period is characterized by a moderately low degree of subsidence with accumulated sediments thickness in the range of hundreds to 1–2 km. The eroded strata thickness is mild, ranging from tens to hundreds of meters during millions of years of erosion. Such small scale uplift and subsidence indicates the basins' dynamic adjustments. Taking the Songliao Basin as an example, the Sifangtai Formation and the Mingshui Formation developed during the subsidence stage are thousands of meters thick, while the Mingshui Formation of Paleogene developed during in the end of the sedimentation phase have very limited thickness and the basin was entirely reversed after that. In the Bohai Bay Basin, after the uplift and erosion by the end of deposition of the Dongying Formation, the basin entered a new phase of overall shrinkage stage when the Guantao and Minghuazhen formations were developed. Another brief uplift and erosion occurred by the end of the Minghuazhen Formation deposition after about 1000 m of sediments were deposited.

② Another important episode for hydrocarbon accumulation and finalizing the petroliferous basin evolution

Despite a moderate rate of subsidence and sedimentation, thousands of meters of sediments may be deposited in the overall shrinkage stage of the basin pushing all major source rocks to the threshold depth for hydrocarbon generation, resulting in an apparent increase in hydrocarbon production. The accompanying brief uplift and erosion renders this phase of development another important episode for hydrocarbon accumulation. In the consequent overall shrinkage or even entire reversal of the petroliferous basins, hydrocarbon accumulations gradually become finalized. For example, after the sedimentation and erosion of the Nenjiang Formation in the Songliao Basin, Sifang and Mingshui formations were deposited, resulting in burial of all major source rocks to the threshold depth. Accompanying the brief uplift and erosion by the end of the Minghuazhen Formation, the main sags and depressions in the Bohai Bay Basin underwent another important episode for hydrocarbon accumulations in the Neogene.

In summary, major source rocks reach the maturity threshold during the overall shrinkage stage and the moderate uplift triggers another hydrocarbon accumulation period. More hydrocarbon charges lead to the formation of batches of new hydrocarbon accumulations. This evolution stage finalizes hydrocarbon accumulations in petroliferous basins.

1.4.4 Implications of Basin Formation, Hydrocarbon Generation and Accumulation Theory

Hydrocarbon generation, migration, accumulation and destruction result from physical and chemical reactions in sediments during basin formation development and shrinkage processes. Hydrocarbons are dynamically involved in basin evolution during subsidence, uplift and shrinkage phases. Therefore, basin formation and its dynamic evolution govern the whole processes of petroleum system. The subsidence stage is a process of material accretion, energy buildup and transformation. Accompanied by physical and chemical interaction under a high temperature and pressure regime, this process fulfills the hydrocarbon generation. The uplift of petroliferous basins is interpreted as energy release as well as hydrocarbon accumulation. The physical and chemical characteristics of this phase are sediments unloading and pressure releasing. The shrinkage stage can be regarded as an energy adjustment and material equilibrium as well as hydrocarbon generation and accumulation optimization. The physical and chemical characteristics involved in this stage are material (energy) equilibrium. In this sense, the implication of petroleum geology is to analyze energy accumulation, release and equilibrium at different stages of petroliferous basin evolution.

Basin formation, hydrocarbon generation and accumulation are intimately inter-associated and complementary, which can be summarized as 'basin formation forms the base, hydrocarbon generation is crucial and hydrocarbon accumulation makes exploration targets real'.

① Basin formation forms the base of petroleum systems. This statement is based on the concept that a basin is the fundamental unit for hydrocarbon occurrence. However, our research will not put too much effort into the morphological classification and genetic mechanisms of basins but focus on petroleum geological evolution characteristics in the stages of subsidence, uplift and shrinkage to characterize the accompanying physical and chemical reactions and their effects on hydrocarbon generation, accumulation and preservation. Therefore, basin formation forms the base of petroleum system and will be regarded as main issue of petroleum geological theories.

② Hydrocarbon generation is critical for the overall process. This idea is based on the decisive role of required fundamental elements for hydrocarbon generation in prospect assessment of sedimentary basins. The subsidence stage is when sediments accumulate in the basin and energy accumulation and transformation under multiple physical and chemical reactions under an increased temperature and pressure regime. Organic matter is converted into hydrocarbons during this process.

③ Accumulated hydrocarbons make exploration success real. This idea is based on the crucial role of hydrocarbon accumulation in petroleum geology theory and exploration practices. The uplifting and erosion is a gradual release of various energy accumulated in the subsidence stage, with the physical and chemical characteristics of energy release and equilibrium under decreased temperature and pressure regime. Hydrocarbon accumulation is merely the manifestation of energy conversions, which will continue to the overall shrinkage stage.

The theory of basin formation, hydrocarbon generation and accumulation is compatible with Magoon's observation in petroleum system. The difference is we not only emphasis the essential elements and processes themselves, but put them all together in basin evolution phases. During the subsidence stage, sediments deposition and organic material accretion form the fundamental base of hydrocarbon generation. During the uplift stage, energy release triggers large scale of hydrocarbon migration and accumulation in structural and stratigraphic traps, while the shrinkage stage makes subsequent migration and more accumulation. The way of thinking in our investigation simplifies complicated petroleum geological issues. We regard a sedimentary basin as one unit to deal with physical and chemical processes involved in basin evolution. Our comprehensive results are more pertaining to mobile resources such as oil and gas.

Chapter 2
Current Status of Hydrocarbon Generation from Source Rocks Theory

Source rock is the most important component of a petroleum system, which directly affects the evaluation of resource potential in the sedimentary basins. Source rock characterization is the foundation of petroleum exploration in all basin research. There are two approaches to characterize hydrocarbon generation from source rocks. One is the kerogen thermal degradation and hydrocarbon generation theory proposed by Tissot and Welte (1978) based on organic geochemistry approach; the other one is called hydrocarbon generation from pore space limited source rocks as proposed in this book and based on a consideration of petroleum geochemistry. The similarities and differences between these two theories are comprehensively summarized in this chapter.

2.1 Kerogen Thermal Degradation and Hydrocarbon Generation Theory

Kerogen thermal degradation and hydrocarbon generation theory was proposed by organic geochemists Tissot and Welte (1978) in their pioneer publication *Hydrocarbon Formation and Occurrence*. This book has a comprehensive discussion of thermal stress constraints on hydrocarbon generation, and a systematic introduction of pyrolysis technologies and their applications to petroleum exploration and resource assessments. This unprecedented work about hydrocarbon generation gained tremendous popularity among organic geochemists, teachers, and students upon its emergence. The authors came to China at the same year when the book was published and gave a copy of this book to Chinese geochemists as a gift. Their ideas were delivered in several lectures and seminars in China during their visit. Their book was undoubtedly timely for Chinese scholars, who were desperate for access to international scientific achievements. In 1982, the Chinese version of this book was issued nationwide. Not long after that their ideas

© Petroleum Industry Press and Springer Science+Business Media Singapore 2017
D. Guan et al., *Theory and Practice of Hydrocarbon Generation within Space-Limited Source Rocks*, Springer Geology, DOI 10.1007/978-981-10-2407-8_2

and research methods were widely applied in petroleum exploration practices in China. Until now, this theory still dominates in domestic and overseas petroleum geochemistry domain.

2.1.1 An Inspiration from Industrial Pyrolysis of Oil Shale

The earliest oil shale retorting to produce shale oil goes back to the seventeenth century, when crude oil was scarce and the petroleum industry relied on oil shale processing to generate man-made oil. The first industrial plant was developed in Autun, France in 1838. Since then many European countries and other countries rich in oil shale such as Brazil and Australia established oil shale retorting plants in the next 100 years. By 1937, shale oil production had reached 500,000 tons/year in Western Europe alone. Meanwhile, developments of the oil shale industry in Europe boosted research on organic geochemistry theories and methods. Starting with oil shale in Scotland, researchers gradually gained the knowledge of organic matter in oil shale and the name of kerogen was created. Kerogen refers to insoluble organic matter which cannot dissolve in organic solvents but can generate oil after heating to 500 °C and higher, while bitumen refers to soluble organic matter which is oil-like material from oil shale. Later, more comprehensive research was conducted concerning kerogen properties, types and characterization methods, giving birth to the discipline of organic geochemistry. After World War II, the flourishing of petroleum exploration resulted in the production of low cost crude oil. Shale oil facilities were successively bankrupted. Despite the gradual decline of oil shale industry, European organic geochemists were inspired from oil shale thermal cracking and shale oil generation, pondered on the issue of hydrocarbon generation from source rocks, and ultimately put forward the theory of kerogen thermal cracking and hydrocarbon generation. This largely accounted for the facts that most petroleum geological theories were proposed by American scholars except for the organic geochemistry theories which were put forward by European scholars. In *Petroleum Formation and Occurrence,* the authors gave this description: the organic matter contained in the oil shale is mainly an insoluble solid material, kerogen. There is no oil and little extractable bitumen naturally present in the rock. Shale oil is generated during pyrolysis, a treatment that consists of heating the rock to ca. 500 °C. The kerogen in oil shale has no distinct difference from the kerogen in petroleum source rocks. Although naturally existing oil in oil shale is negligible, it can generate shale oil upon the process of pyrolysis. Similarly, source rocks can generate oil by elevated temperature upon burial. These two processes are comparable and methods developed from oil shale retorting may be applicable to source rocks thermal evolution assessment.

Based on this theoretical consideration, a simple experimental simulation of source rocks-derived kerogen evolution could reveal the whole natural transformation of kerogen. Therefore, further research on experimental simulation of source rocks thermal cracking was conducted by organic geochemists.

1. Studies on kerogen

Figure 2.1 perfectly demonstrates how organic matter evolved into kerogen and geochemical fossils in source rock sedimentation and diagenesis processes.

Based on solvent extraction, elemental analysis, and physical and chemical property assessment, kerogens have been comprehensively investigated. Three types of kerogens have been identified and their thermal evolution paths have been tracked.

2. Experimental simulations of kerogen thermal evolution

Experiment assays of thermal evolution have been carried out on kerogen for shallow immature samples, which are applied to the naturally transformed samples

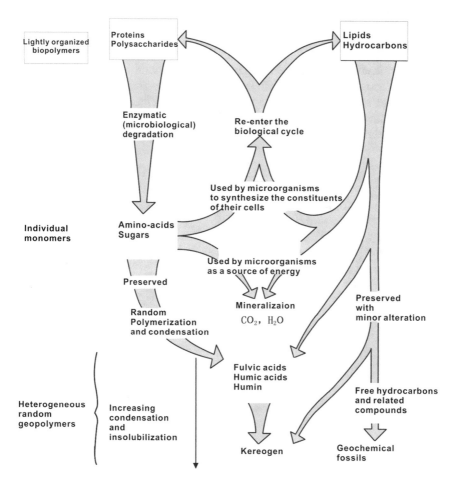

Fig. 2.1 Fate of organic matter during sedimentation and diagenesis, resulting in two main organic fractions: kerogen and geochemical fossils (after Tissot and Welte 1978)

at greater depths from the same basin. Three successive stages are classified based on laboratory heating experiments of immature source rocks, i.e., diagenesis stage—the source rocks are immature and the products of this stage are CO_2, water, nitrogen, sulfur and oxygen compounds; catagenesis stage—kerogen encounters its major degradation stage, with the products being crude oil and gas condensate; metamorphism stage—most of the product is dry gas and the residue is dead carbon. Based on these observations, the general process of oil and gas generation from kerogen is summarized in Fig. 2.2.

At the same time, the laboratory heating of source rocks suggested that under rapidly increasing temperature, properties of kerogen in experimental simulations are comparable to those of naturally transformed kerogen under geological heating conditions. This discovery perfectly verified the application of experimental data in accounting for properties of kerogen buried at greater depths in every stage of kerogen thermal degradation.

3. Assessment criteria and identification techniques of source rocks

The experimental simulation of kerogen thermal evolution makes the transformation from kerogen to oil accessible in a laboratory. By analyzing the outcome of experimental simulations of source rocks, samples at different burial depth (temperature) with different organic matter types and contents provide assessment criteria and identification techniques of source rocks. The main ideas are as follows:

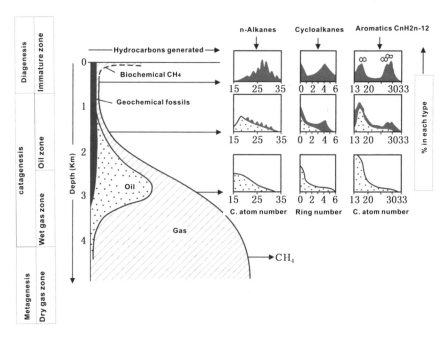

Fig. 2.2 General scheme of hydrocarbon formation as a function of burial of source rocks (after Tissot and Welte 1978)

i. Amount of organic matter

The minimum TOC content of source rocks is 0.3 % for carbonate rocks and 0.5 % for mudstones.

ii. Maturity of organic matter

Four successive stages of kerogen evolution can be distinguished according to vitrinite reflectance (%Ro) values along the evolution pathway: below 0.5 %Ro (sometimes 0.7 %Ro) is the diagenesis stage and organic matter is immature; between 0.5 and 1.3 %Ro is the catagenesis stage and organic matter matures with most oil generation at this stage; between 1.3 and 2.0 %Ro is the late catagenesis stage and organic matter is highly matured with light oil and gas condensate as the main products; beyond 2.0 %Ro is the metagenesis stage and organic matter is overmature with dry gas as the only product.

iii. Pyrolysis methods and assessment of source rock samples

These pyrolysis experiments normally heat 100 mg sample with the temperature programmed to 500 °C under an inert atmosphere in an attempt to represent a progressive burial. During the heating process, weight loss is first caused by the release of free and adsorbed hydrocarbons, which is determined by flame ionization detector as S_1. The S_1 peak is detected during the initial temperature while it is held for 3 min at 300 °C. Then the weight loss is successively caused by pyrolytic hydrocarbon and hydrocarbon-like compounds (S_2), and oxygen-containing volatiles like CO_2 and water. The last parameter equals to the maximum temperature (T_{max}) under which hydrocarbon amount reaches the maximum rate of generation during artificial evolution (Fig. 2.3).

This artificial evolution in heating experiments can be applied for semiquantification assessment of source rock potential of hydrocarbon. S_1, the original generation potential of source rocks, represents the petroleum (oil and gas) already formed by the kerogen; S_2, the residual generation potential, represents the petroleum which is about to be generated from the same kerogen; thus $S_1 + S_2$ is the estimated total amount of petroleum that the kerogen is capable of generating, i.e., the total generation potential. For example, rock with a total generation potential below 2 kg/t is poor in oil generation but has the potential to generate gas. Medium quality source rock has a total generation potential between 2 and 6 kg/t; for excellent quality source rocks, the value would be greater than 6 kg/t.

2.1.2 *From Destructive Distillation Oven of Oil Shale to Rock-Eval*

Since the theoretical consideration of this theory is inspired by the pyrolysis of oil shale to generate shale oil, the simplest way to obtain oil and gas from source rocks

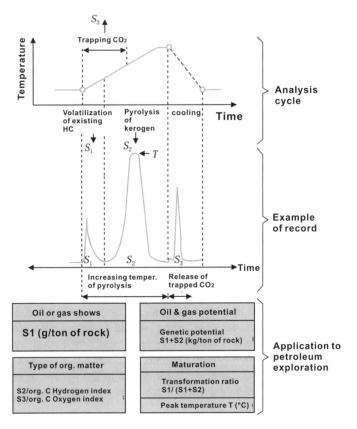

Fig. 2.3 Cycle of analysis and example of record obtained by the pyrolysis method of Espitalié et al. (1975)

is by developing experimental simulation devices similar to industrial retorting technology for extracting oil from oil shale.

1. Destructive distillation of oil shale is a process under which oil shale samples are heated under inert atmosphere in an attempt to degrade the organic matter into shale oil. The temperature under which generation of shale oil begins is generally termed as the initial temperature of pyrolysis, and the temperature at which the maximum shale oil generation rate is called peak temperature of pyrolysis, and the final temperature of shale oil generation is called final temperature of pyrolysis (Table 2.1).

  Three distinct stages in oil shale thermal retorting can be identified according to temperature ranges as follows:

① Desiccation of oil shale: from ca. 0–350 °C, when temperature is below 100 °C, water inside oil shale evaporates; CO_2 and water vapor are generated. When the temperature reaches the range of 180–200 °C, oil shale goes through the process of deoxidation, generating bitumen, water, and gas.

Table 2.1 Temperature range of pyrolysis of major source rocks in China (Shi Guoquan 2009)

Origin of shale	Temperature (°C)			
	Beginning pyrolysis temperature	Final pyrolysis temperature	Range of pyrolysis	Peak pyrolysis temperature
Fushun	403	507	104	470
Maoming Jintang	396	517	121	–
Maoming Yangjiao	347	550	203	455
The 4th member of Huadian	418	513	95	456
The 11th member of Huadian	411	530	119	470
Yilan of Heilongjiang	408	527	119	480
Yongdeng of Gansu	430	504	74	470
Tianchi of Xinjiang	356	497	141	–
Longkou of Shangdong	356	510	154	470

② The major degradation stage of oil shale: When the temperature reaches the range of 350–500 °C, the already existing pyrolytic bitumen will be further degraded and into shale oil, water, coke, and gas in great amount.

③ The final temperature phase: When the temperature reaches the range of 500–550 °C, it is well beyond the peak temperature of oil shale pyrolysis, with shale oil production rate being the lowest.

A thermal retorting device for extracting oil from oil shale has a very simple structure. Taking domestic Fushun distillation oven as an example: this oven is composed of a sample supply device, oven body, and ash treatment device. The oven body is the main part, of which the upper distillation segment and the lower reaction segment are separated by an arch plate in the middle (Fig. 2.4).

In industrial practice, the first step is adding oil shale (about 10–75 mm diameter pieces) to the oven from the top, and then it is dried in the thermal retorting segment by the upward flow of gas from the bottom (temperature is 500 °C), with the generated shale oil and gas escaping from the oven top. The oil shale organic matter is transformed into coke and falls into the reaction segment, where it is blended with the air and water vapor vented from the upper part entering from the oven bottom, burn to oil shale ash and finally come out through the ash vessel at the bottom. The generated gas of burning reactions goes upward and acts as the heating gas in the thermal retorting part. Another part of heating comes from the combustion of the cooled and condensed thermal retorted gas in the heat preservation oven. These two heat sources have improved heating efficiency of the thermal retorting technology.

Fig. 2.4 Structure illustration of Fushun destructive distillation oven (Qian 2006)

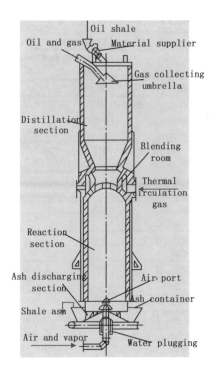

2. The Rock-Eval

To address the problem of time-consuming and expensive pretreatment to obtain kerogen, Espitalié developed a new device–Rock-Eval (Fig. 2.5), to simulate the oil shale destructive distillation device. Rock-Eval is capable of assessing source rocks directly using core or cuttings.

The Rock-Eval apparatus heats source rock samples to 550 °C in an inert atmosphere. When the temperature rises from room temperature to 300 °C, the S_1 will first be liberated. When temperature goes up from 300 to 500 °C, S_2, S_3 (CO_2) and oxygenic volatiles like water are generated successively; when temperature goes up to 500–550 °C, dry gas is finally generated. At this stage, all kerogen in source rock samples has been exhausted. Four fundamental parameters (S_1, S_2, S_3, and the temperature at maximum yield of pyrolysis (T_{max}), which are closely associated with source rocks properties) are recorded. Source rock assessment can be conducted by an analysis of them. Specifically, $S_1 + S_2$ (mg/g, hydrocarbon/ rock) suggests source rocks potential, $S_1/(S_1 + S_2)$ is an indication of oil production index of source rocks, S_2/S_3 represents source rock type index.

3. Comparison between Rock-Eval and retorting technology of oil shale

Rock-Eval is a laboratory simulation of natural transformation of a trace amount of kerogen, while thermal retorting technology of oil shale is for massive production of shale oil in industry. Both of them are thermal treatment of source rocks to investigate kerogen degradation. They have similar pyrolyzing conditions,

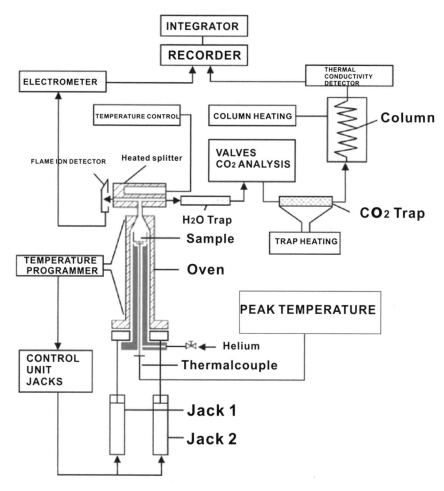

Fig. 2.5   Principles of the Rock-Eval pyrolysis device by Espitalié et al. (1975)

temperature programing, products and measuring units but the differences between them are also significant (Table 2.2).

The industrial practice of shale oil production by oil shale thermal retorting was published as early as the English Patent 330 of Martin Al in 1694, which described detailed approaches to extracting a great amount of bitumen, tar, and oil products. Present thermal retorting devices are improved versions for industrial production. Therefore, the artificial transformation of oil shale is simple: desiccate the exploited oil shale in preprocesses and then heats it in the thermal retorting device, which are in essence organic geochemistry reactions. Regardless of the fact that kerogen transformation for hydrocarbon generation is accessible to us in laboratory simulation by the Rock-Eval, many controlling factors other than temperature are ignored. Temperature is very important but by no means the only concern in the

Table 2.2 Contrast between oil shale pyrolysis and source rock pyrolysis

Item	Oil shale retorting	Rock-Eval
Pyrolysis condition	Heated to 500 °C under isolation air	Heated to 500 °C under inert gas
Changes under pyrolysis temperature	0–350 °C: Dehydration starts at temperatures below 120 °C to produce CO_2 and steam. When temperature reaches 200 °C and above, small amount of soluble bitumen, water and gas are produced. 350–500 °C: main shale oil production stage together with large amount of coke. 500–550 °C: the final pyrolysis temperature	0–350 °C: Produce CO_2 and steam at temperatures below 200 °C, produce free hydrocarbons at temperature 200–350 °C (S_1); produce crude oil, wet gas (S_2), CO_2 (S_3) and other volatiles containing oxygen at temperature 350–500 °C; 500–550 °C, produce dry gas
Product of pyrolysis	Shale oil, gas, semicoke, shale ash, water, CO_2 etc.	Extractable bitumen (S_1), crude oil and wet gas (S_2), dry gas, volatiles containing oxygen, water and CO_2
Unit of measurement	Kg/t, ratio of shale oil and oil shale, final yield of shale oil	potential of oil generation $S_1 + S_2$: kg/t (hydrocarbon/source rocks)

simulation of complicated natural transformation of kerogen. As we all know, kerogen natural transformation in source rocks is accompanied by source rock diagenesis, during which changes in source rock mineralogy, connate water, crystal water, and static pressure of burial depth will affect directly the kerogen transformation. In conclusion, these laboratory simulations fail to take geological conditions into consideration although they could perfectly demonstrate kerogen evolution and its properties.

To sum up, Rock-Eval recommend by Tissot, which is developed by Espitalié, is a simulation of industrial thermal retorting technology of oil shale. The experimental simulations with this device undoubtedly fail to account for the multiple variables involved in natural processes, making it nothing more than organic chemistry experiments. The application of experimental data obtained by this device in evaluating natural source rocks will definitely generate misleading results.

2.1.3 Hydrocarbon Yield Versus Source Rock Generation Potential

The book *Petroleum Formation and Occurrence* gives a detailed description of a mathematical model, a quantitative approach to evaluate oil and gas prospects. The contents include mathematical models of kerogen degradation and hydrocarbon generation, generation potential of source rocks, transformation ratio, and their applications in petroleum exploration. In summary, the main ideas are quantification of generated hydrocarbon and locating petroleum prospects.

1. Hydrocarbon quantification

According to the mathematical model developed by Tissot and Espitalié for experimental simulations of kerogen evolution in basin, the computational formula of hydrocarbon generated (quantification of hydrocarbon generated within only a certain evolution stage) is as follows:

$$Q = 10^{-8}S \cdot h \cdot \rho \cdot Cr \cdot Rc \cdot Hr$$

In the formula

Q is overall hydrocarbon yield, t;
S is area of effective source rocks, km^2;
h is average thickness of source rocks, m;
ρ is source rocks density, t/km^3;
Cr is residual TOC of source rocks, %;
Rc is recovery coefficient;
Hr is hydrocarbon productivity of source rocks in this phase of evolution, kg/t (hydrocarbon/TOC).

When Hr refers liquid hydrocarbon productivity, then Q refers to petroleum production; when Hr refers to gaseous hydrocarbon productivity, then Q refers gas production (Zhao 1999).

The six parameters in this computational formula express different components of hydrocarbon. $S \cdot h \cdot \rho$ equals to total weight of source rocks (t); $Cr \cdot Rc$ equals to the original total TOC content by percentage (%); Hr refers petroleum (gas) productivity of source rocks in this stage of evolution (kg/t). $S \cdot h \cdot \rho \cdot Cr \cdot Rc$ refers original TOC by weight; $S \cdot h \cdot \rho \cdot Cr \cdot Rc \cdot Hr$ is the total amount of oil or gas that source rocks is capable of generating at the transformation ratio of this evolution stage. This is an organic chemistry formula of reaction, the burning of materials on right side of formula generating hydrocarbon on the left under the assumption that organic matter on the right side of the formula are completely converted at that transformation ratio. Obviously, the result obtained by this formula is a source rock generation potential rather than actual hydrocarbon yield. Based on this assumption, no potential remains once source rocks reach a certain evolution stage and organic carbon will become graphite. However, the fact is, most source rocks are not exhausted at thermal evolution stage proposed by Tissot and their TOC contents remain high. For example, residual TOC content of source rocks in 1st member of Qingshankou Formation, Songliao Basin is 2.21 %; residual TOC content of source rocks in upper part of 4th member of Shahejie Formation, Jiyang sag, Bohai Bay basin is 2.24 %. Original total TOC of these two formation are 3.98 and 4.03 %, respectively, with TOC recovery coefficient taking a value of 1.8 (TOC recovery coefficient of type I kerogen is between 1.5 and 2.0 by experience). In these two samples, the ratio of residual TOC to original total TOC is around 56 %, which indicates that the depleted TOC in pyrolysis is 44 % of the original total TOC. As a conclusion, a large portion of original total TOC has been preserved in source rocks in the form of residual TOC instead of being involved in the reaction of hydrocarbon generation. This phenomenon so commonly occurs that organic geochemists like B.P. Tissot are supposed to be aware of it. Question is, why do they still treat source rocks generation potential as hydrocarbon production? The reason may lie in them equaling source rocks to oil shale all the way. Just as B.P. Tissot has described in *Petroleum Formation and Occurrence*, 'to some extent the pyrolysis leading to shale oil formation from kerogen is comparable to the burial of the source rocks at depth that generates oil by resulting elevation of temperature because shallow burial and immature oil shale, containing sufficient amount of organic matter shows no substantial difference from source rocks'. Organic rich, shallow buried immature sediment is oil shale, while deeply buried oil shale constitutes petroleum source rock. When oil shale is heated to around 500 °C, all organic matter could have been thermally cracked for shale oil and gas formation; when source rocks is heated to 550 °C by using Rock-Eval, the rest of the organic matter will be completed cracked to generate oil and gas. Thermal history is the only difference between oil shale and source rocks. Once heated, all residual organic matter in source rocks will largely transform into hydrocarbon as long as temperature and time permit. This may have explained why B.P. Tissot regarded source rocks potential as hydrocarbon yield.

2. Locating oil and gas source rock potential in a basin (depression and sag)

The mathematical model of kerogen degradation and hydrocarbon generation proposed by Tissot and Espitalié was based on kinetics of kerogen degradation and used the general scheme of evolution. Since kerogen is a macromolecule composed of polycondensed nuclei bearing alkyl chains and functional groups, the links between nuclei being heteroatomic bonds or carbon chains, the bonds are successively broken roughly in the order of increasing rupture energy, as the burial depth and temperature increase. The products generated are first heavy heteroatomic compounds, carbon dioxide, and water; then progressively smaller molecules; and finally hydrocarbons (Fig. 2.6).

The model provides a computation of the amount of oil and gas generated and migrated in any place of the basin as a function of time. Data required are:

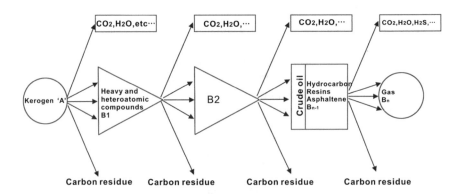

Fig. 2.6 General scheme of kerogen degradation (Tissot and Espitalié 1975)

exploration and production well locations in the basin (depression and sag), a map of depositional facies, a stratigraphic isopach map of the main formations, thickness isopach maps of the main strata, dark mudstone (shale), TOC content, Ro value of source beds, kerogen type distribution pattern, and hydrocarbon productivity curves of pay beds. Based on those maps and indices, hydrocarbon generation potential in the overall source rocks area is finally obtained. Generally, the quantity of hydrocarbon yields can be measured by g/kg (petroleum/TOC) and g/t (petroleum/rock) (Figs. 2.7 and 2.8). In the study of hydrocarbon generation history by basin modeling, oil and gas generation intensity are generally expressed by hydrocarbon generation potential in source rocks areas, measuring by 10^4 t/km^2 (for oil) and 10^8 m^3/km^2 (for gas) (Fig. 2.9).

This method of mathematical model or computation formula demonstrates system reactions of kerogen degradation under the temperature compensating for time system in source rocks samples. Only a limited number of steps are considered in particular (Fig. 2.10). Eventually, kerogen in source rocks has been transformed into hydrocarbons and carbon residue, in other words, 'no transformation at shallow burial depth and completely transformed once deeply buried'. As a result, the ultimate outcome of this model is hydrocarbon generation potential of source rocks, rather than actual hydrocarbon yields in a certain evolution stage.

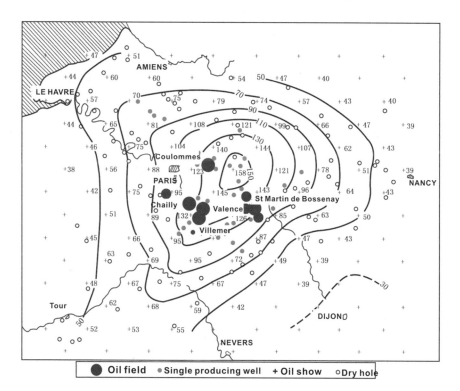

Fig. 2.7 Application of the mathematical model to the generation of oil from Lower Jurassic source rocks of the Paris Basin. The oil generated is expressed in g petroleum per kg organic carbon (after Tissot and Welte 1978)

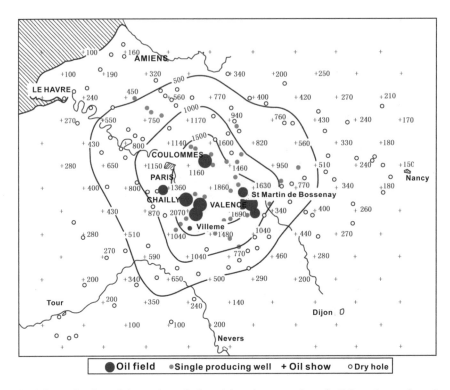

Fig. 2.8 Application of the mathematical model to the generation of oil from Lower Jurassic source rocks of the Paris Basin. The oil generated is expressed in g petroleum per t rock (Tissot and Welte 1978)

2.1.4 An Analysis of the Theory of Kerogen Evolution and Hydrocarbon Generation

This theory consists of three parts. The first is kerogen formation and evolution characteristics. Based on the organic origin of petroleum, this theory describes detailed composition, deposition, preservation, and accumulation of organic matter; analyzes systematically geochemical characteristics of lipid, protein, carbohydrates, and lignin the organic matter components, and microbial degradation, condensation effects, and kerogen evolutional pattern in ways other than dissolution; and introduces concretely kerogen compositions, classification and definition methods. This part is featured by an unprecedented proposal of the natural evolution of organic matter to oil and gas with kerogen as the intermediate, and formed the foundation of contemporary theory of petroleum organic origin. The second part is a comprehensive elaboration of mechanisms of kerogen thermal evolution to generate oil and further crack to gas. By analysis of kerogen diagenesis, catagenesis and metamorphism of organic matter, and experimental simulations of kerogen evolution, the theory summarizes the compete process of kerogen transformation to oil, gas

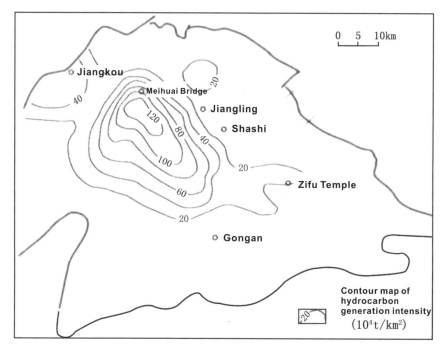

Fig. 2.9 Hydrocarbon generation intensity of Xinxia Member source rocks in the Jiangling Sag, Jianghan Basin (after Luo et al. 2013)

Fig. 2.10 Successive oil and gas formation from kerogen

condensate and dry gas. It outlines temperature related origin of petroleum, which forms the main idea of this theory. The third part is application of this theory in petroleum exploration practices, which is generally source rock evaluation and petroleum prospect evaluation of petroliferous basins. For example, the frequently applied computational simulation approaches are developed on the basis of mathematical simulations in recent years. In addition, there is an introduction about standard procedure for laboratory assays of source rocks by Rock-Eval as well as analysis techniques of items as source rocks organic matter type, content, identification of maturity, parameters selection of oil, and source rocks correlations.

The first part of this theory is fundamental tenets that should be mastered by petroleum geochemists and petroleum geologists. This is the most valuable part both for academic research and practical application. However, the second and third parts are unjustifiable in that source rock evolution study is an indiscriminating simulation of the industrial practices of oil shale pyrolysis and shale oil generation, on a presumption that kerogen in shale oil and source rocks show no essential difference to each other. Problems will definitely exist in this theory in terms of both the way of thinking and its application in petroleum exploration.

The foremost flaw will be misrepresentation of natural geological conditions in the organic chemistry approach instead of petroleum geochemistry approach.

The theory of source rock degradation and hydrocarbon generation is the main issue of petroleum geochemistry. Since petroleum geochemistry is a combination of petroleum geology and chemistry (specifically organic chemistry), this issue should be addressed with petroleum geochemistry way of thinking, i.e. the integrated analysis of physical and chemical conditions of source rock evolution. Organic matter within source rocks is formed in a sedimentary basin (depression), which coexists with other minerals and thermal evolution of organic matter has become inseparable parts (source rock integration) interacting with each other. However, this theory based on oil shale retorting and Rock-Eval pyrolysis isolates source rock thermal evolution from a petroleum geochemical way of thinking which emphasizes interaction within a basin. The practice of this theory turns complicated petroleum geochemical problems into simple organic chemical ones. There is no fault from the perspective of organic chemical reactions and the theory is supported by organic chemistry rules. However, using organic chemistry approach instead of petroleum geochemistry one is surely a flaw of this theory as natural geological conditions are overlooked.

The appropriate way is to integrate source rock sedimentary evolution with synchronous kerogen thermal evolution based on rules of petroleum geochemistry. According to sedimentology principles, in the last stage of sedimentation of overburden strata, muddy sediments containing organic matter have filled the whole space of the basin (depression). When source rocks are buried to the threshold depth, space for chemical reactions (also the space to hold petroleum) is pores in the source rocks at this stage of sedimentation. As burial depth increases, source rock porosity decreases under the compaction effects of overburden rocks, while hydrocarbon has been generated all the time. This process of 'decrease and increase' will certainly result in filling of source rock pore space in a certain stage of evolution, when oil, gas, and water reach a state of relative equilibrium, kerogen thermal degradation is restricted, the untransformed organic matter remains in the source rocks, under the constraints of pore space and pressure of over burden formations. Furthermore, experimental simulations should be designed on the basis of the petroleum geochemistry ways of thinking (detailed experimental design is in the second section of this chapter). This experiment design should better resemble the actual situation of hydrocarbon generation in source rocks under geological conditions.

Second, petroleum prospect evaluation of petroliferous basins proposed by this theory is hydrocarbon potential generation quantity, which may never be reached.

Kerogen thermal degradation models proposed by this theory and currently widely applied in computational basin models regard source rocks as solid minerals as oil shale and compute the total amount of oil and gas generated according to chemical reaction formulas, supposing that kerogen has been exhausted after thermal evolution. To get a clear understanding of this idea, the computation formula of oil shale transforming to shale oil and that of source rocks to hydrocarbon are compared here.

- Computational formula of shale oil generation by oil shale

$$Q_{\text{shale}} = S \times H \times D \times W$$

In this formula

Q_{shale} refers generation quantity of shale oil, 10^4 t;
S refers area of oil shale, km^2;
H refers thickness of oil shale, m;
D refers density of oil shale, t/km^3;
W refers oil content of oil shale, %.

- Computational formula of petroleum generation by source rocks

In the formula: $Q = 10^8 \times S \times h \times \rho \times Cr \times Rc \times Hr$
Q refers overall hydrocarbon yield, t;
S refers area of effective source rocks, km^2;
h refers average thickness of effective source rocks, m;
ρ refers source rocks density, t/km^3;
Cr refers residual TOC of source rocks, %;
Rc refers recovery coefficient;
Hr refers hydrocarbon productivity of source rocks in this stage of evolution, kg/t (hydrocarbon/TOC).

The two formulas follow the same rule of material balance in organic chemistry reactions. By the computation outcome of petroleum amount generated by oil shale or source rocks, oil generation can be easily assessed by transformation ratio and the source rock is exhausted by the end of evolution. This thought is correct for oil shale as the computation amount of shale oil can be generated by the industrial practice of shale oil thermal retorting. However, it is untenable when applied in hydrocarbon generation by source rocks during natural evolution as it is a dynamic process of petroleum geological chemistry, in which there are dynamic equilibrium and interactions of oil, gas, and water under joint functioning effects rather than mere temperature. Organic matter (kerogen) in source rocks is impossible to transform to 100 % transformation ratio, on this account, the hydrocarbon generation yield obtained by this organic chemical way of thinking is only a potential or theoretical amount that is unlikely in natural evolution. The theory applies this unrealistic hydrocarbon yield as actual hydrocarbon yield in petroleum resource evaluation and adopts artificial parameter options like 'expulsion coefficient' and 'accumulation coefficient' to compute hydrocarbon expulsion and accumulation amount. This will definitely mislead petroleum resource evaluation and exploration, and result in randomness and artificiality.

Third, to regard the source rocks the same as oil shale causes confusion in source rock assessment.

This theory does not regard properly the essential differences between kerogen in source rocks and oil shale. Oil shale retorting to generate oil differs significantly from source rocks generating oil and gas upon burial.

Evaluation criteria proposed in this theory can be briefly summarized as follows:

TOC content: Minimum TOC content in source rocks of mudstone is 0.5 % and that of carbonate rock type is 0.3 %. However, there's no upper limit, with obviously the more, the better.

Kerogen type: There are three types of kerogen: sapropel, humic and mixed. Generally the sapropel type is referred to as type I, the humic as type III and mixed as type II. Type I is oil prone; type III is gas prone; type II is in between. In terms of oil generation, apparently type I is most favorable.

Kerogen maturity: Measured by light reflectance in humic and vitrinite components, four obvious thermal evolution stages can be identified.

$Ro < 0.5$–0.7 %, the stage of diagenesis. Source rocks are immature;

0.5–0.7 % $< Ro < 1.3$ %, the stage of catagenesis. This is the main interval for oil generation (oil window);

1.3 % $< Ro < 2.0$ %, the stage of late catagenesis. This is the main interval for wet gas and condensate oil generation;

$Ro > 2.0$ %, the stage of metamorphism. This is interval for dry gas generation.

Kerogen capacity of hydrocarbon generation (expressed by kg organic matter per t rock):

Lower than 2 kg/t, doesn't count source rocks and has only limited capacity of gas generation;

2–6 kg/t, fair quality source rocks; above 6 kg/t, good quality source rocks.

According to the above criterion, organic geochemical indices of oil shale in China reveal unexceptionally good quality source rocks except the maturity with Ro less than 0.6 % (Table 2.3). Based on Tissot's definition of source rocks maturity, source rocks with Ro value under 0.6 % are immature, and thus not source rocks. So he had to rename the oil shale as 'potentially economically valuable sediments containing large quantity of kerogen'.

The question is can oil shale transform into source rocks simply due to time and burial depth? Table 2.3 shows that oil shale of the Permian Lucaogou Formation in the Yaomoshan mountainous region in Xijiang has the highest Ro value around 0.6 % and remains immature. The same aged mudstone becomes mature source rock in the Junggar Basin. Oil shale of the Upper Triassic Yanchang Formation in Tongchuan, Shanxi Province is mature at R_o value of 0.51 %, while the same age mudstone is mature at the same R_o value. Why does oil shale mature later than lacustrine mudstone? The reason is that oil shale under geological conditions has rarely been buried to the depth where temperature is over 150 °C. Such temperature is not sufficient for the degradation of over 10 % of the organic matter to form more than a small amount of soluble bitumen. Oil shale cannot be a source rock in most situations. The difference between Carboniferous oil shale (muddy shale) and

Table 2.3 Organic geochemical characteristics of oil shale in the main Chinese oil shale mining areas (Liu 2009)

Location	R_o (%)	TOC (%)	Peak temperature in pyrolysis T_{max} (°C)	Potential of oil generation $S_1 + S_2$ (mg/g)	Type index S_2/S_3	Hydrogen index S_2/OC (mg/g)	Oxygen index S_3/OC (mg/g)	H/C	O/C
Liandahe of Heilongjiang	0.46	7.481	436	37.59	41.02	488.04	4.07	1.14	0.13
Liandahe of Heilongjiang	0.46	27.45	423	72.4	28.22	257.05	9.11	0.95	0.21
Liandahe of Heilongjiang	0.5	19.66	444	117.02	144.71	588.86	11.9	1.33	0.08
Huadian of Jilin	–	33.305	443	230.21	135.75	680.71	5.01	1.47	0.1
Luozigou of Jilin	0.48	17.96	437	105.97	96.5	542.71	5.62	1.35	0.09
Fushun of Liaoning	0.49	11.72	443	67.73	63.55	563.91	8.87	1.33	0.09
Fushun of Liaoning	–	19.43	446	120	143.05	603.71	4.22	1.48	0.06
Dongsheng of Neimenggu	0.41	38.02	439	208.21	12.78	535.32	41.87	1.28	0.14
Maoming of Guangdong	0.51	15.88	428	82.15	40.39	493.39	12.22	1.32	0.17
Maoming of Guangdong	0.51	24.46	429	128.25	46.24	500.94	10.83	1.34	0.15
Maoming of Guangdong	0.56	29.82	429	153.28	91.31	496.08	5.43	1.3	0.13
Maoming of Guangdong	0.58	24.34	433	129.31	97.38	512.08	5.26	1.31	0.12

(continued)

Table 2.3 (continued)

Location	Ro (%)	TOC (%)	Peak temperature in pyrolysis T_{max} (°C)	Potential of oil generation $S_1 + S_2$ (mg/g)	Type index S_2/S_3	Hydrogen index S_2/OC (mg/g)	Oxygen index S_3/OC (mg/g)	H/C	O/C
Tongchuan of Shanxi	0.51	25.4	432	71.26	12.07		21.26	1.05	0.2
Sangonghe of Xinjiang	0.51	18.91	435	24.73	14.19		9.04	1.25	0.15
Yaomoshan of Xinjiang	0.59	9.251	439	43.3	3.94		117.28	1.24	0.16
Yaomoshan of Xinjiang	0.6	20.265	442	109.75	29.68		17.76	1.41	0.13

Table 2.4 Characteristics of representative oil shales in China (Zhaojun Liu 2009)

ore-bearing area	Color	Density (t/m³)	Organic carbon (%)	Average oil yield (%)	Maximum oil yield (%)	Average ash content (%)	Average calorific value (MJ/kg)
Huadian	Gray brown, yellow brown	1.99	33.31	8.59	24.80	69.37	9.99
Fushun	Brown, yellow brown'	2.12	13.06	5.86	12.00	76.10	4.75
Huangxian	Dark brown, brown black	1.64	–	13.82	18.49	58.45	11.66
Dalianhe	Gray-gray brown	2.00	27.45	6.85	9.12	64.35	7.59
Luozigou	Black-light black	1.94	17.96	6.04	14.37	76.39	9.65
Nongan	Gray-gray brown	2.04	14.33	4.85	12.10	82.55	4.19
Binxian	Black	2.21	12.72	6.20	7.00	79.04	–
Tongchuan	Dark brown	1.90	25.40	6.53	9.25	74.05	–
Maoming	Dark brown, yellow brown	1.80	23.63	6.47	13.00	74.20	6.90
Danzhouchangpo	Gray-dark gray	1.60	–	4.87	–	70.74	5.68
Yaojie	Light black, black	2.07	–	5.55	17.72	69.87	6.93
Yaomoshan	Black, light brown	2.17	14.76	7.00	14.92	77.35	7.85

mudstone is not difficult to distinguish as the former, which has at least 12 % TOC, contains more than 5 % distillable oil while the latter has no requirement (Table 2.4). The difference between oil shale and coal is their ash content as the former has more than 40 % of ash while the latter is less than 40 %. Although the TOC content in coal may well exceed that in oil shale, the H/C ratios are reversed. Organic matter in source rocks is mature and has experienced the process of hydrocarbon generation. As a conclusion, oil shale and source rocks are generally different rock types. In spite of similarities between them, they cannot be regarded as the same without discrimination.

2.2 Theory of Hydrocarbon Generation in Pore Space Limited Source Rocks

This theory was proposed more than ten years ago, but due to lack of laboratory apparatus, experimental simulations could not be conducted. This theory is only available in our book *Theoretical Way of Thinking of Basin Formation, Hydrocarbon Generation and Accumulation*. Since 2004, under the guidance and support of the department of scientific development of Sinopec, the authors and affiliated research groups have spent ten years in independently developing the geological model of hydrocarbon generation and expulsion under in situ temperature and pressure conditions. Based on the practical data of Bohaiwan Basin, Nanxiang Basin and Baiyinzagan Basin, systematic research has been conducted on the theory of hydrocarbon generation in pore space limited source rocks and the preliminary theory of hydrocarbon generation has been developed.

2.2.1 Our Theory Is Derived from Petroleum Geochemistry Consideration

The theory of basin formation, hydrocarbon generation and accumulation in the fourth section of Chapter 1 provides detailed descriptions about the three stages in petroliferous basin evolution, which are 'continuous subsidence stage—material loading and pressure increasing; the general uplift stage—unloading and hydrocarbon accumulation; and the overall shrinkage phase—material equilibrium establishment between the basin and surrounding regions, hydrocarbon reservoir improvement and settlement'. Therefore, the study of hydrocarbon generation in source rocks can only be dealt with by application of petroleum geochemistry approaches, presentation of basin petroleum geological evolution history and analysis and summary of theoretical issues on hydrocarbon generation. Aspects of the theory of hydrocarbon generation in pore space limited source rocks are sketched in Fig. 2.11.

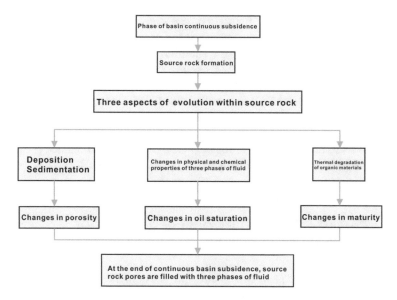

Fig. 2.11 Illustration of the aspects of the theory of hydrocarbon generation in pore space limited source rocks

Research on petroleum geological evolution history of source rocks involves three aspects: The first one is source rock sedimentation–diagenesis process and characteristics by sedimentological principles–focusing on source rock porosity alteration with time. The second one is organic matter thermal evolution and characteristics from the perspective of petroleum geochemistry. The third is phase status and characteristics of oil, gas, and water and their dynamics under continuous subsidence and growing pressure due to sediment loading, by combined methods of physical chemistry and petroleum geology. Apart from these subjects, we have also discussed conceptual issues on space for petroleum geochemical reactions.

As is known to all, no chemical reactions will take place without three basic elements: reactants, reaction conditions (temperature and pressure, for instance) and holding space, i.e., where the reaction takes place (as small as beaker, test tube and as large as industrial devices). Petroleum geochemical reactions within source rocks are no exception. For those reactions, reactants, temperature, and pressure are not difficult to determine, while the subject of reaction space needs additional consideration.

It is common sense from the petroleum geological theories that source rocks occur in reducing and stagnant environments such as sags (or depressions) in sedimentary basins. Their size is large but reaction space is limited. Such conditions take direct control on source rock formation, mineral compositions, organic matter content, distribution and depth range, and water salinity. Accordingly, as source rock has filled up the whole space during sag (or depression) subsidence by the end of a certain sedimentation phase, reaction space for organic matter thermal degradation and holding space for oil, gas, and water can only be provided by pore space in the source rock system formed at the same time. Previous research has shown that, in the continuous basin subsidence stage, source rocks gradually solidify with porosity decreasing and organic matter getting more mature under increasing temperature. When source rock maturity (Ro) increases to 0.5 %, porosity shrinks to 10–20 %, that is to say, the 10–20 % porosity hosts the whole geochemical reaction in which source rock maturity (Ro) changes from 0.5 % to the degree of hydrocarbon generation. As a result, either on a macroscale or a microscale, both source rocks formation and degradation take place in limited pore space, which has significant impact on hydrocarbon generation in such system.

2.2.2 Fundamental Characteristics of Source Rock Petroleum Geological Evolution

From the perspective of petroleum geology, it is a complicated research subject to study source rock diagenesis, which involves many facets. This book focuses on porosity alteration, maturation evolution and their relationships, which are closely related to hydrocarbon generation from source rocks. To make it clear, we applied practical data of Dongying Sag in the analysis.

1. Source rock porosity alteration

A correlation chart between porosity and burial depth is based on measured data of effective source rocks from different depths in Dongying Sag and previously published porosity evolution in shales and mudstones (Fig. 2.12).

This graph shows that, as burial depth of effective source rocks in Dongying Sag increases, its porosity changes with the following notable characteristics: ① A zone of rapid porosity decrease. Compaction intensifies rapidly to a depth of about 100 m. Porosity decreases from an initial value of 50–60 to 30 %, and loosely accumulated muddy sediments turn to weakly consolidated aggregates, owing to the expulsion of pore water by the load of over burden sediments. ② Moderate porosity decrease zone. This change takes place when sediments are buried to less than 900 m, where porosity decreases from 30 to 20 %, weakly consolidated sediments become semi-consolidated. Clay minerals are largely composed of smectite. The decrease in porosity is closely related to the expulsion of pore water. ③ A transitional zone of porosity change. The change takes place at a burial depth of 900–1500 m, where porosity alters only slightly from 20 to 16.5 %. It is probably because free water has been exhausted from the sediments, with the rest in the form of bound water in illite/smectite disordered mixed layers transformed from smectite. Muddy sediments change from semi-consolidated to consolidated mudstone or shale. ④ Zone of slowly decreasing porosity. This change takes place at a depth of 1500–4200 m, porosity gradually decreases from 16.5 to 5.2 %, which largely owes to the transformation of disordered illite/smectite mixed layers to

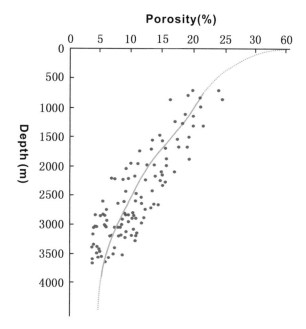

Fig. 2.12 Correlation between porosity and depth of effective source rocks in Dongying Sag

ordered illite/smectite layers and illite, resulting in the transformation of bound water to free water and then the expulsion of free water under compaction. Organic matter in this stage has thermally evolved from immature to mature. The existence of organic acid to a certain degree promotes the transformations of clay minerals to illite and of bound water to free water. Meanwhile, generated oil will initially fill the pores. Within the oil window, oil saturation in pores gradually goes up until oil and water are relatively saturated. Without any force triggering oil expulsion, compaction of overburden gives rise to abnormal pressure within source rocks. There is no substantial change in porosity. Hydrocarbon generation is restricted. ⑤ A zone of porosity stability. This zone is buried at a depth greater than 4200 m, where porosity shows no substantial change and remains around 5 %.

Porosity evolution during sedimentation and lithification has direct effects on kerogen evolution and hydrocarbon generation. Judging from Fig. 2.13, burial depth of source rocks in the oil window in basins with high geothermal gradient is far less than that in low geothermal gradient basins. A shallowly buried source rock has a larger relative porosity in which more hydrocarbons can be held. This accounts for the fact that Songliao Basin has abundant petroleum resources and Miyang Depression has small areas but abundant and quality petroleum resources.

1. The determination of organic matter maturity in source rocks

One of the significant parameters in source rock assessment is organic matter maturity. Ro value shows obvious signs of being restricted because of the dominance of Type I and Type II$_1$ organic matter in the main source rocks of Dongying Sag.

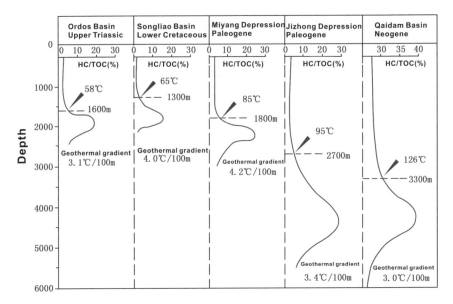

Fig. 2.13 Curves of transformation ratio of organic carbon to hydrocarbon in several basins in China

To accurately determine the main source rock maturation of Dongying Sag, we applied FAMM technique to measure maturation. Maturity obtained by the FAMM technique is expressed as equivalent vitrinite reflectance. Samples for FAMM analysis are collected in the main source rocks of the middle–lower part of the Es$_3$ member and the upper part of the Es$_4$ member. Geological and geochemical characteristics of samples are shown in Table 2.5, as well as analysis results by Guo (2003). The table shows samples including mudstone and shale, buried at a depth of about 2024.2–3905.2 m. TOC (total organic carbon) contents of samples are all above 1.0 %, mostly above 1.5 %, with a few exceptions and several above 4.0 %. Tmax is between 433 and 458 °C, generally above 435 °C. Hydrogen index (I$_H$) is between 145 and 779 mg/g, largely above 300 mg/g. Organic matter types of samples include type I, II$_1$, II$_2$ and III, mainly type I and I$_1$. The measured %Ro value is between 0.3 and 0.85 %. FAMM analysis show fundamental characteristics of the Dongying Sag main source rocks and are shown in Fig. 2.14 and Table 2.6.

Based on the fluorescence variation diagram of representative source rocks of different types (Fig. 2.14), all source rock samples deviated from the normal vitrinite reflectance calibration curve except a few type III ones. Type I kerogen is generally close to 0.30 % calibration curve; Type II$_1$ generally falls near the 0.20 % calibration curve and Type II$_2$ generally falls near the 0.10 % calibration curve. This suggests that vitrinite reflectance has been depressed to varying degrees for different types of source rocks in Dongying Sag. It seems that the better the quality of source rocks, the more the vitrinite reflectance has been depressed. A comparison of equivalent vitrinite reflectance (EqVRo) obtained by FAMM and measured vitrinite

Table 2.5 Fundamental characteristics of samples from Dongying Sag in FAMM analysis

Well name	Lithology	Depth (m)	Age	TOC (%)	T_{max} (C)	I_H (mg/g)	Organic type	VR_o (%)	Origins of data
Liang225	Mudstone	2240.5	Es_4	1.68	438	436	II_1	0.41	This project
Liang242	Mudstone	2433.3	Es_3	1.91	437	479	II_1	0.54	
Lai108	Shale	2479	Es_3	3.24	435	779	I	0.36	
Ying93	Mudstone	2562.4	Es_3	1.68	435	358	II_1	0.43	
Bo11	Mudstone	2593	Es_4	2.21	440	376	II_1	0.60	
Niu5	Mudstone	2598	Es_4	4.25	442	675	I	0.48	
Chun371	Shale	2757.8	Es_3	1.89	437	466	II_1	0.55	
Bin417	Mudstone	2844	Es_4	2.26	441	520	I	0.42	
Ying93	Shale	2865.2	Es_3	5.91	442	691	I	0.42	
Liang242	Mudstone	2921.8	Es_3	1.97	436	269	II_1	0.57	
He88	Mudstone	3050	Es_3	1.87	441	416	II_1	0.54	
Niu33	Mudstone	3133	Es_3	2.65	441	560	I	0.53	
Ying921	Mudstone	3159.1	Es_3	2.67	441	546	I	0.58	
Ying891	Mudstone	3187.6	Es_3	1.98	445	410	II_1	0.52	
Wang54	Shale	3241.4	Es_3	5.66	444	669		0.50	
Shi122	Mudstone	3402.2	Es_3	1.25	442	359	II_1	0.63	
Wang57	Shale	3423.2	Es_3	5.83	442	576	I	0.6	
Fengshen1	Mudstone	3686.6	Es_4	1.7	446	288	II_1	0.67	
Wang78	Shale	3732.6	Es_3	2.21	443	283	II_1	0.61	
Lai64	Mudstone	3795	Es_4	1.41	444	253	II_1	0.68	
Wang78	Shale	3905.2	Es_3	1.93	445	247	II_1	0.85	
Liang225	Mudstone	2024.2	Es_4	0.56	435	145	III	0.52	
NIu38	Shale	2790	Es_4	1.07	440	158	III	0.62	
L38	Shale	2805	Es_4	1.02	434	384	II_1	0.49	
L38	Shale	3046	Es_4	1.6	433	362	II_1	0.54	
L38	Mudstone	3188	Es_3	2.95	440	535	II_1	0.58	
L38	Mudstone	3253	Es_3	1.91	440	380	II_1	0.64	
L38	Mudstone	3310	Es_3	1.51	437	384	II_1	0.53	Guo et al. (2003)
T73	Mudstone	2497	Es_3	1.66	435	241	II_2	0.47	
T73	Mudstone	2893	Es_3	2.04	437	412	II_1	0.46	
T73	Mudstone	2994	Es_3	2.56	441	432	II_1	0.50	
T73	Mudstone	3377	Es_3	2.08	444	304	II_2	0.64	
T73	Mudstone	3403	Es_3	2.18	448	411	II_1	0.67	
Y182	Mudstone	2506	Es_3	1.54	436	264	II_2	0.53	
W7	Mudstone	2630	Es_3	4.01	439	616	I	0.40	
W35	Mudstone	2172	Es_3	4.34	438	637	I	0.30	
W128	Mudstone	3731	Es_3	2.95	458	161	I	0.78	

reflectance (VRo) of 37 source rocks samples shows that the difference is distinct. The measured vitrinite reflectance in type I source rocks has been underestimated in the range of 0.24–0.35 % with an average value of 0.30 %. For type II_1, it is in the

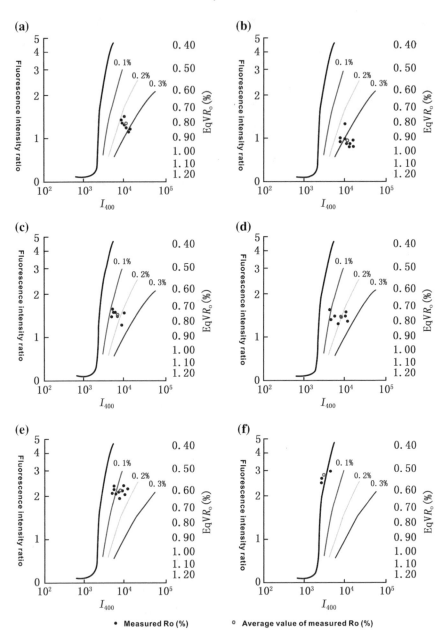

Fig. 2.14 Fluorescence alteration illustration of different types of source rocks in Dongying Sag. **a** Sample 54564, Wang54, 3241.4 m, Type I, EqVRo = 0.81 %; **b** sample 54566, Wang57, 3423.2 m, Type I, EqVRo = 0.92 %; **c** sample 67593, Bo11, 2593 m, Type II$_1$, EqVRo = 0.75 %; **d** sample 54557, Chun 371, 2757.8 m, Type II$_1$, EqVRo = 0.78 %; **e** sample S192, T73, 2497 m, Type II$_2$, EqVRo = 0.60 % (Data from Guo et al. 2003); **f** Liang225, 2024.2 m, type III, EqVRo = 0.54 %

Table 2.6 Comparison between measured Ro and equivalent Ro (EqVRo) in FAMM analysis of source rocks samples of Donging Depression

Well name	Depth (m)	Lithology	Organic type	VR_o (%)	$EqVR_o$ (%)	$EqVR_o$-VR_o (%)	Origins of data
Liang225	2240.5	Mudstone	II_1	0.41	0.61	0.2	This project
Liang242	2433.3	Mudstone	II_1	0.54	0.71	0.17	
Lai108	2479	Shale	I	0.36	0.68	0.32	
Ying93	2562.4	Mudstone	II_1	0.43	0.62	0.19	
Bo11	2593	Mudstone	II_1	0.6	0.75	0.15	
NIu5	2598	Mudstone	I	0.48	0.77	0.29	
Chun371	2757.8	Shale	II_1	0.55	0.78	0.23	
Bin417	2844	Mudstone	I	0.42	0.73	0.31	
Ying93	2865.16	Shale	I	0.42	0.73	0.31	
Liang242	2921.8	Mudstone	II_1	0.57	0.73	0.16	
He88	3050	Mudstone	II_1	0.54	0.73	0.19	
Niu33	3133	Mudstone	I	0.53	0.77	0.24	
Ying921	3159.06	Mudstone	I	0.58	0.79	0.21	
Ying891	3187.6	Mudstone	II_1	0.52	0.8	0.28	
Wang54	3241.4	Shale	I	0.5	0.81	0.31	
Shi122	3402.2	Mudstone	II_1	0.63	0.85	0.22	
Wang57	3423.22	Shale	I	0.6	0.92	0.32	
Fengshen1	3686.6	Mudstone	II_1	0.67	0.94	0.27	
Wang78	3732.57	Shale	II_1	0.61	0.83	0.22	
Lai64	3795	Mudstone	II_1	0.68	0.97	0.29	
Wang78	3905.2	Shale	II_1	0.85	0.98	0.13	
Liang225	2024.2	Mudstone	III	0.52	0.54	0.02	
Niu38	2790	Mudstone	III	0.62	0.65	0.03	
L38	2805	Mudstone	II_1	0.49	0.66	0.17	Rutai et al. (2003)
L38	3046	Mudstone		0.54	0.74	0.2	
L38	3188	Mudstone	II_1	0.58	0.79	0.21	
L38	3253	Mudstone	II_1	0.64	0.82	0.18	
L38	3310	Mudstone	II_1	0.53	0.85	0.32	
T73	2497	Mudstone	II_2	0.47	0.6	0.13	
T73	2893	Mudstone	II_1	0.46	0.68	0.22	
T73	2994	Mudstone	II_1	0.5	0.8	0.3	
T73	3377	Mudstone	II_2	0.64	0.82	0.18	
T73	3403	Mudstone	II_1	0.67	0.86	0.19	
Y182	2506	Mudstone	II_2	0.53	0.68	0.15	
W7	2630	Shale	I	0.4	0.76	0.36	
W35	2172	Shale	I	0.3	0.65	0.35	
W128	3731	Shale	I	0.78	1.08	0.3	

Fig. 2.15 Diagram of equivalent Ro value in effective source rocks versus burial depth in Dongying Sag

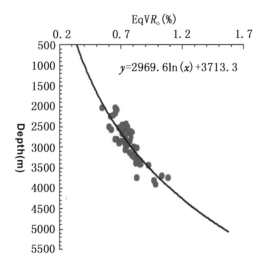

range of 0.16–0.30 % with an average value of 0.22 %. For type II_2, it is in the range of 0.13–0.18 %, with an average value 0.15 %. However, almost no depression occurs in type III kerogen and the difference between two methods is about 0.02 %.

Figure 2.15 shows correlations between samples measured using FAMM methods and their corresponding burial depths. Source rocks in Dongying Sag currently buried at a depth of about 1500 m have Ro value of about 0.5 % and are marginally mature. Some source rocks currently buried at a depth of 3100 m have Ro value of around 0.80 % and are in the peak oil generation stage. Those currently buried at 4400 m are highly mature.

According to the correlation between porosity and burial depth of source rocks in Dongying Sag (Fig. 2.12), quantitative relationships between porosity and maturity of effective source rocks can be determined (Table 2.7, Fig. 2.16). Using these results, a model of hydrocarbon generation in pore space limited source rocks in Dongying Sag has been established. Table 2.7 and Fig. 2.16 show that, when calibrated Ro of effective source rocks vitrinite reflectance equals to 0.5 %, its porosity is 16.5 % and the corresponding burial depth is around 1500 m; when

Table 2.7 Correlation between porosity, depth, and Ro (calibrated) of effective source rocks samples of Dongying Sag

R_o (calibrated) (%)	Porosity	Corresponding burial depth
0.50	17.0	1500
0.60	11.5	2200
0.70	8.5	2700
0.80	7.5	3100
0.90	6	3400
1.00	5.5	3750
1.10	5.2	4000
1.20	5.0	4200

Fig. 2.16 Correlation (after calibrations) between porosity and Ro in effective source rocks in Dongying Sag

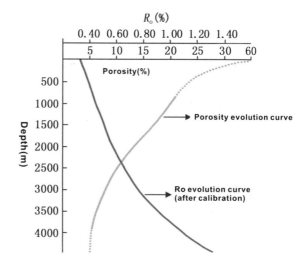

calibrated Ro is 1.2 %, porosity decreases to 5.0 %, and corresponding depth is 4200 m. In conclusion, pore space evolution and source rocks thermal evolution are simultaneously occurring during source rock catagenesis. They are closely inter-related as catagenesis is intensified, porosity reduces continuously and kerogen becomes more mature progressively.

2.2.3 Experimental Simulations of Hydrocarbon Generation in Pore Space Limited Source Rocks

1. Apparatus to simulate hydrocarbon generation and expulsion under in situ conditions

Both geological models of hydrocarbon generation and expulsion in source rocks and experimental designs should abide by petroleum geochemical principles and approach, i.e., the natural process of hydrocarbon generation and expulsion of source rocks in geological conditions.

i. A brief introduction of the device

The DK modeling device of hydrocarbon generation and expulsion under in situ conditions is shown in Fig. 2.17, which basically consists of six parts: high temperature and pressure reaction system, bidirectional hydraulic pressure control system, hydrocarbon expulsion (primary migration) system, automatic control and data collection system, product separation and collection system, and peripheral support systems.

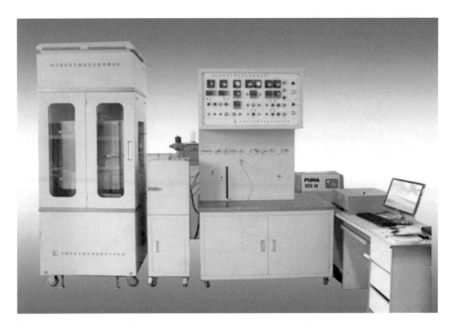

Fig. 2.17 The DK modeling device of hydrocarbon generation and expulsion under in situ conditions

The high temperature and pressure reaction system is basically composed of a high temperature and pressure reaction container, various sealing elements, high-temperature electric oven, connecting valves and pipes. The high temperature and pressure reaction container (Fig. 2.18) is the core of the whole device, where samples are placed, experimental simulations are conducted and various products accumulate. As the reaction container must withstand high temperature and pressure and the strong corrosion of the media, the critical techniques are its

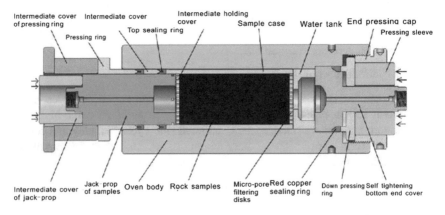

Fig. 2.18 Construction diagrams of high temperature and pressure reaction container

temperature and pressure resistance (safety), ability to avoid oxidation or reduction (corrosion resistance) and resistance to the likely strong corrosion and extreme tightness. This container is made of the latest domestic alloy steel of high strength, which is capable of standing a maximum stress of 350 MPa at 600 °C, 8 times ordinary stainless steel ($1Cr_{18}Ni_9Ti$). The yield strength and tensile strength of this alloy steel far exceed those of $1Cr_{18}Ni_9Ti$, so it is unlikely to generate creep under high temperature (600 °C) and high pressure (100 MPa) and the consequent micro leakage of the barrel, thus guarantee safety during experiments.

The mandatory hot air circulation and high temperature electrothermal box shaped furnace mainly consists of a stainless steel lining, replaceable iron, and nickel metal resistance wire, insulating layers made of special ceramic fiber material, preheating and interlayer heating circulation system, mandatory blowing-in and high-temperature velocity shifting engine, high-temperature hot air circulation entry and preheating cavity. The heating oven has the following characteristics: ① high speed heating from room temperature to 1000 °C with adjustable rates in only 15–30 min. ② The programmer allows up to eight heating ramps with intermediate temperature holds to automatically maintain temperature ramping and temperature consistency. It also has the functions of a PID parameter self-setting, manual/automatic shifting without interruption and temperature excursion alarming. ③ High accuracy of the temperature control can be achieved with small overshoot (less than 2 °C), temperature homogeneity of ±15 °C, accuracy of ±1 °C. The maximum design temperature is 800 °C.

The hydrocarbon expulsion system (primary migration) includes a high pressure hydrocarbon expulsion container, high pressure balancing pump, and gas promoted electric valves for hydrocarbon expulsion and fluid supplement, reaction container and connecting pipes for the collecting devices. The maximum controllable pressure of gas promoted electric valves for hydrocarbon expulsion and fluids supplement can be as high as 180 MPa, effective pressure can be adjusted anywhere between normal pressure and 120 MPa, either manually or automatically. The application of hydrocarbon expulsion valves can have generated hydrocarbon expelled under programmed fluid pressure, thus achieving the goal of controllable hydrocarbon generation and expulsion (episodic hydrocarbon expulsion), as well as a geological model of hydrocarbon generation within a controlled pressure range. This valve is also a safety valve, whose control signals are connected with the alarm and discharging fluids to reduce pressure before the pressure of the reaction systems exceeds the maximum operating pressure of the devices. Gas promoted electric valves for fluid supplement function to provide fluid pressure balance when the system pressure decreases because of fluid reactions during hydrocarbon generation and hydrocarbon expulsion and during fixed fluid pressure experimental simulations. In this modeling experiment, top and bottom ports are simultaneously used for hydrocarbon expulsion and fluid supplementation. The main consideration for such a design is to mimic the compacted cores under high pressure as the generated and expelled fluids will be expelled in either direction. If expulsion occurs only in one side, it is very likely that ultrahigh fluid pressure will occur at the opposite side, causing danger and more importantly, inconsistency between experimental

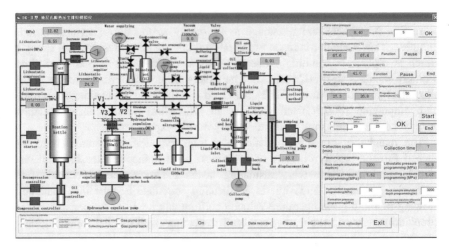

Fig. 2.19 Operation interface of automatic control

hydrocarbon generation and geological conditions. The maintenance of differential pressure is obtained by the connection of the hydrocarbon generation system to an expelled hydrocarbon collection system, which is capable of maintaining an extremely high pressure (maximum operating pressure is 150 MPa).

The automatic control and data collection system (Fig. 2.19) consists of a pressure control system and fluid entry and exit alarm control system. The main components are high accuracy heater, high-temperature hydraulic pressure sensor, intelligent temperature controller, various pressure indicators and one computer. The heating and control are achieved by temperature transmitter and intelligent temperature controller. A heating and pressure modeling experiment is closely related to the accuracy of the temperature program for heated samples and temperature control accuracy, i.e., the accuracy of temperature measurement and control of the heater. The core component of temperature control is heat sensors, whose accuracy is ±(0.15 + 0.2 % FS) and liner range is 0–800 °C, meeting the national level A standard, with an error range of ±0.2 °C. Lithostatic pressure and formation fluid pressure are displayed and controlled by high temperature hydraulic pressure sensors. The design of this system fully takes into consideration the lithostatic pressure and formation fluid pressure, making them more similar to actual geological conditions. The pressure range is between 0 and 200 MPa, with an accuracy of ±0.2 MPa. The data collection system is mainly composed of data collection cards, computers, printers, and their supporting equipments. Data collected include mechanical pressure (lithostatic pressure), fluid pressure (formation pressure), simulation time, temperature, and sample numbers, all of which can be acquired in real time with data collecting cards and computers.

The product separation and collection system is composed of an automated collector, water cooling circulator, products (gases and liquids) gauging device, and vacuum pump. The gauging device requires a high degree of accuracy to get accurate quantities of gases and liquids (especially light hydrocarbons). The peripheral assistance system includes high-pressure manual and electric metering pumps, a high-pressure intermediate container, a preparation device of cores for the experiments, a platform for sample loading and unloading, and other tools. The device shell consists of a frame, control panel, and carbon steel shell with painted plastic. The frame provides support for the whole device. It consists of one top and one bottom iron plate for standing pressure, four stainless steel pillars for standing pressure, four omnidirectional wheels and one foundation. To elevate lithostatic pressure of the sample chamber, hydraulic pressure work stations can reach 45×10^4 N, and the thickness of the top and bottom plates can be 30 cm.

ii. Chief technical parameters and technical characteristics of the modeling device of hydrocarbon generation and expulsion under in situ conditions

(1) Chief technical parameters

① Simulation temperature: the maximum operating temperature is 600 °C. Heating, holding and cooling can be realized through programming, with an accuracy of ±2 °C, an actual temperature deviation within 3 °C during a continuous 24 work hours.

② Pressure: Lithostatic pressure and fluid pressure can be programmed simultaneously with this device. The maximum simulation value of overburden formation pressure (directly imposed on the rock) can be equal to lithostatic pressure at a burial depth of 8000 m, which is 200 MPa; the maximum simulation value of fluid pressure can be as high as the fluid pressure deep in basins at about 10 km, which is 100 MPa. Additionally, 0–80 MPa of fluid pressure can be obtained in the hydrocarbon expulsion chamber.

③ The sample chamber has an internal diameter of 25–38 mm, a height of 50–80 mm, a maximum sample weight of 150 g, which can meet ordinary requirements for geochemistry analysis.

④ Heat and pressure modeling methods for hydrocarbon generation are chiefly models under high pressure in totally enclosed environments, models of fluid continuity in semi-enclosed environments (hydrocarbon generation accompanied by expulsion/intermittent expulsion) and models for hydrocarbon generation in open environments (continuous hydrocarbon expulsion).

⑤ Major ways of hydrocarbon expulsion include differential pressure hydrocarbon expulsion, hydrocarbon expulsion caused by compaction, one-off products expulsion, and continuous hydrocarbon expulsion during generation.

⑥ Product collection and accurate metering, gas-flow rate computation (flow meter and automatic gas collection by water displacement), gas and fluid product condensation (controllable mechanical/electronic condensation with the lowest temperature of −20 °C), and collection installations. Liquid products are weighed quantitatively, while gaseous products are quantified by means of gas chromatography. The residual oil products in hydrocarbon generation and expulsion systems are extracted by mixed solution of high pressure organic solutions and supercritical CO_2.

⑦ Automatic control, display and recording of time, temperature, lithostatic and fluid pressure, and gas quantity

(2) Technical characteristics

This device is a controllable system of hydrocarbon generation and expulsion. It maintains a source rock's original mineral compositions and occurrence states and similar pore space as in situ conditions for geochemical reactions where liquid water of high pressure (formation liquid) is filled up. It provides natural simulation conditions of lithostatic pressure, formation pressure, and confining pressure where organic matter degrades for hydrocarbon generation and expulsion in high temperature and within limited time.

(3) Fundamental functions

① Lithostatic pressure from overburden strata and confining pressure (a maximum value of 200 MPa) can be imposed simultaneously with higher formation pressure (100 MPa), under which hydrocarbon generation and expulsion from source rocks can be conducted.

② Original core or artificially compacted core samples are adopted to maintain original pores and fabric as far as possible and to guarantee the sample has the original states of organic matter and formation pore space where hydrocarbon generation and expulsion take place.

③ Hydrocarbon generation and expulsion can not only be conducted under controlled fluid pressure in airtight conditions but also by means of episodic hydrocarbon expulsion, achieved by controllable differential pressure between the source and storage chambers.

④ Water is preserved in the form of liquid in rock sample pores, making the device a real hydrous hydrocarbon generation and expulsion model.

⑤ Hydrocarbon generation and expulsion efficiency can be simulated with source rocks of different types, under different temperature and pressure, in different fluid media and different inorganic mineral settings.

⑥ The following experimental conditions can be optionally programmed or automatically adjusted: experimental duration, temperature, sampling amount, lithostatic pressure, formation pressure, and fluid characteristics, ways of hydrocarbon expulsion and pressure difference used to allow hydrocarbon expulsion.

2. Experimental approach to hydrocarbon generation and expulsion simulation

i. Selection of samples

For a systematic study of hydrocarbon generation and expulsion of source rocks in Chinese continental rift basins, we picked 10 source rocks samples of low maturity, different TOC content and different rock types out from a plenty of samples to conduct experimental simulations of hydrocarbon generation and expulsion under in situ conditions. Samples are largely from Miyang Depression, Dongpu Depression, Baiyinchagan Sag and Huadian outcrop in Jilin Province. Geochemical characteristics of those samples are listed in Table 2.8.

ii. Experimental methods

(1) Sample preparation and experiment setup

Taking source rock heterogeneity into consideration, we ground the samples into 40–60 mesh, mixed and then divided them into several portions. For each programmed temperature, the sample portion was compressed in small cylinder shape to guarantee uniformity and representativeness. The detailed sample loading pattern can be adapted according to requirements (Fig. 2.20).

① Pure shale: Simulation was conducted with small cylinder core made of only source rock (60–100 g). ② Sand–shale stacked: The bottom of the small cylinder core is source rocks (60–100 g) and the top of the core is sandstone (20–40 g). ③ Shale–sand stacked: The top of the small cylinder core is source rocks (60–100 g) and the bottom of the core is sandstone (20–40 g). ④ Sand–shale–sand sandwiched: Both the top and the bottom of the small cylinder core are sandstones (20–40 g), and the middle is source rocks (60–100 g). ⑤ Shale–sand–shale sandwiched: Both of the top and the bottom of the small cylinder core are source rocks (60–100 g), and the middle is sandstone (20–40 g).

(2) Programed temperature and pressure conditions in experiments

The major consideration of temperature and pressure in experiments is to match in situ conditions such as burial depth (corresponding relationships between Ro and burial depth), static pressure and formation pressure. Ten groups of temperature and pressure programming in hydrocarbon generation and expulsion simulating models are shown in Table 2.9.

(3) Experimental procedure

① Leakage test: Install core sample into the reaction container, compress it, seal it, inject inert gas until the pressure reaches 5–10 MPa, then allow to stand and observe any pressure decline. If it is completely tight, release the inert gas and finally, repeatedly evacuate the device and inject gas into it 3–5 times, and finally close the reaction container under vacuum.

② Water injection: Inject 60–80 MPa water with a high pressure pump (pure water, saline water, or formation water) may be used to fill up the pores in the compressed core sample. (The core sample will absorb water resulting in a continuous decrease in fluid pressure, so decrease stable pressure indicates the saturation of pores with water). To ensure the whole hydrocarbon generation and expulsion system is filled with water, fluid pressure cannot be less than 2–3 MPa before heating.

Table 2.8 Basic geochemical characteristics of samples for experimental simulations

Sample number	Sample location	Lithology	Depth (m)	Age	R_o (%)	TOC (%)	Chloroform bitumen "A" (%)	S_1 (mg/g)	S_2 (mg/g)	I_H (mg/g)	Organic type
Huadian-8	Huadian	Gray mudstone		Palaeogene	0.36	6.40	0.0714	0.07	36.98	578	II_1
Wang-24	Miyang Depression	Black mudstone	1271.3	Eh_3	0.55	4.55	0.4209	0.74	36.399	763	I
Wei-20	Dongpu Depression	Salt-bearing shale	2290.0	Eh_3	0.62	3.93	0.5359	0.88	25.56	592	II_1
Cha-1	Baiyinchagan Basin	Dark gray mudstone	1788.8	K_1bd_3	0.56	3.45	0.3720	0.51	17.41	505	II_1
Bi-215	Miyang Depression	Gray mudstone	1532.6	Eh_3	0.66	2.56	0.0738	0.08	15.83	600	II_1
Huadian-3	Huadian	Gray mudstone		Paleogene	0.42	2.29	0.0265	0.01	10.66	466	II_1
ZY-M	Peizhi				0.56	1.87		0.27	7.43	397	II_1
Pu-1-154	Dongpu Depression	Breen mudstone	2364.0	Es_1	0.63	1.30	0.2844	0.21	5.87	376	II_1
Huadian-6	Huadian	Gray mudstone		Paleogene	0.46	0.89	0.0099	0.00	1.16	130	II_2
ZY-L	Peizhi				0.56	0.78		0.07	1.55	199	III

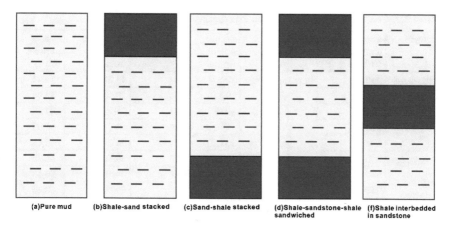

(a)Pure mud   (b)Shale-sand stacked   (c)Sand-shale stacked   (d)Shale-sandstone-shale sandwiched   (f)Shale interbedded in sandstone

Fig. 2.20 Examples of sample loading patterns

③ Simulations of compaction and heating: Use the hydraulic press to program the pressure to the value under which the sample was compacted. Start the temperature controller and heat the sample to the programmed value at a rate of 1 °C/min, and then hold the temperature constant for 48–96 h.

④ Simulation of hydrocarbon expulsion: For a continuous subsidence simulation, sustain pressure equivalence between the hydrocarbon generation system and the hydrocarbon expulsion system. To simulate episodic hydrocarbon expulsion, open the exit valve for hydrocarbon expulsion once the pressure of the hydrocarbon expulsion system has exceeded that of the hydrocarbon generation system by some amount before rapidly closing it. These operations are repeated throughout the process of hydrocarbon generation. Because hydrocarbon generation and expulsion models operate in a sealed environment, open the exit valve for hydrocarbon expulsion at an early phase of temperature rise and close it to prevent hydrocarbon expulsion when the pressure of the hydrocarbon generation system has reached formation pressure for a certain depth.

(4) Product collection and quantification

① Gas collection and quantification: Connect the product collection and quantification device and hydrocarbon expulsion device to the system, then collect gases from the hydrocarbon generation system and hydrocarbon expulsion device to model hydrocarbon generation and expulsion under differential pressure in a continuous subsidence environment, and to model hydrocarbon generation and expulsion in a sealed environment, collect gases from the hydrocarbon generation system.

When the whole reaction system cools to 150 °C, open the valve for hydrocarbon expulsion to release oil, gas, and water from the hydrocarbon generation and expulsion system (Fig. 2.21).

The first fluid expelled is a mixture of water, gas, and light oil dissolved in gas, which is separated after going through liquid collecting pipes and cooled using

Table 2.9 Condition setups in simulation experiments

Sample	Simulated temperature (°C)	Sample arrangement	Sample weight (g)	R_o (measured) (%)	Lithostatic pressure (MPa)	Fluid pressure (MPa)
Huadian-8	250	Pure shale	102.71	0.44	46.00	52.44
	275		100.86	0.54	50.60	32.94
	300		103.67	0.61	52.90	47.71
	310		104.28	0.67	54.05	48.99
	320		103.20	0.71	55.20	46.21
	330		104.89	0.78	59.80	48.38
	340		104.55	0.86	64.40	50.13
	350		103.67	1.06	69.00	52.16
	360		103.13	1.24	73.60	49.21
	370		104.89	1.67	78.20	66.10
	380		101.13	1.85	82.80	64.89
	400		104.27	2.10	88.55	53.46
Wang-24	250	Pure shale	60.89	0.35	52.80	43.0
	300		60.69	0.44	63.60	47.0
	320		60.42	0.54	69.60	54.7
	340		60.69	0.64	81.60	61.9
	350		60.69	0.72	84.50	66.5
	360		60.08	0.89	88.80	75.7
	375		60.37	1.22	102.00	101.9
	385		60.78	1.64	112.00	117.9
	400		80.92	2.20	135.00	121.0
Wei-20	275	Sand/shale/sand	60.92	0.44	46.00	65.6
	300		60.51	0.52	50.60	24.4
	315		60.93	0.55	52.90	26.6
	330		60.85	0.71	55.20	26.9
	340		60.21	0.82	59.80	35.4
	350		60.95	1.14	64.40	35.0
	360		60.08	1.25	69.00	38.6
	370		60.75	1.63	73.60	40.5
	380		60.13	1.68	78.20	44.1
	400		60.50	2.11	82.80	45.3
Cha-1	250	Sand/shale/sand	60.98	0.62	36.8	23.3
	275		60.10	0.63	39.1	24.9
	300		60.75	0.72	43.7	30.7
	310		60.55	0.80	46	29.8
	320		59.14	0.89	50.6	31.4
	335		58.28	1.07	55.2	41.9
	350		59.85	1.28	59.8	41.2
	360		59.48	1.45	64.4	40.4
	370		58.60	1.76	73.6	47.8
	380		60.32	1.93	80.5	51.4
	400		60.49	2.06	92	61.6

(continued)

Table 2.9 (continued)

Sample	Simulated temperature (°C)	Sample arrangement	Sample weight (g)	R_o (measured) (%)	Lithostatic pressure (MPa)	Fluid pressure (MPa)
Bi-215	275	Sand/shale/sand	59.16	0.35	52.8	
	300		60.98	0.41	63.6	34.7
	325		59.97	0.53	69.6	36.6
	350		60.02	0.81	81.6	38.2
	360		60.15	0.98	88.8	44.4
	370		60.12	1.52	91.2	45.2
	380		60.88	1.67	96	46.5
	400		60.90	2.06	105	52.7
Huadian-3	250	Pure shale	100.58		46.00	33.8
	275		100.77		50.60	44.9
	300		101.40		52.90	41.8
	310		102.12		54.05	47.5
	320		100.55		55.20	49.7
	330		100.82		59.80	53.8
	340		102.42		64.40	55.6
	350		102.54		69.00	62.9
	360		101.50		73.60	65.6
	370		101.59		78.20	51.3
	380		100.12		82.80	67.7
	400		101.82		88.55	63.1
ZY-M	275	Pure shale	62.73	0.91	34.50	40.3
	300		62.20	1.07	39.10	49.2
	310		61.56	1.29	43.70	34.9
	320		61.81	1.35	46.00	35.4
	330		61.78	1.43	47.00	40.8
	340		61.82	1.48	49.45	42.5
	350		60.14	1.53	52.00	63.8
	360		61.18	1.64	55.20	47.1
	370		60.48	1.93	57.50	46.6
	380		60.16	2.08	64.40	52.1
	400		60.73	2.26	69.00	46.4
Pu-1-154	250	Pure shale	60.02	0.63	52.80	35.7
	300		60.46	0.71	63.60	47.0
	320		60.21	0.79	69.60	50.8
	340		60.43	0.87	74.40	59.3
	350		60.02	0.95	81.60	58.7
	360		60.30	1.35	88.80	59.3
	375		60.00	1.44	91.20	58.7
	385		60.08	1.49	96.00	63.2
	400		60.11	1.95	105.00	73.2

(continued)

Table 2.9 (continued)

Sample	Simulated temperature (°C)	Sample arrangement	Sample weight (g)	R_o (measured) (%)	Lithostatic pressure (MPa)	Fluid pressure (MPa)
Huadian-6	250	Pure shale	102.45	0.57	46.00	76.8
	275		102.16	0.68	50.60	41.03
	300		102.98	0.75	52.90	27.24
	310		102.22	0.91	54.05	43.22
	320		103.37	0.95	55.20	26.94
	330		106.36	1.05	59.80	43.08
	340		101.80	1.15	64.40	40.81
	350		103.24	1.19	69.00	56.94
	360		105.80	1.27	73.60	37.14
	370		107.41	1.53	78.20	52.68
	380		106.10	1.84	82.80	54.57
	400		100.80	2.1	88.55	54.27
ZY-L	275	Pure shale	60.12	0.75	34.50	70.92
	300		60.51	1.11	39.10	
	310		60.55	1.2	43.70	30.3
	320		62.16	1.26	46.00	43.7
	330		61.41	1.37	47.00	36.6
	340		62.13	1.43	49.45	41.9
	350		59.96	1.48	52.00	49.4
	360		60.16	1.69	55.20	56.5
	375		60.67	1.91	57.50	53.5
	400		60.15	2.21	64.40	48.3

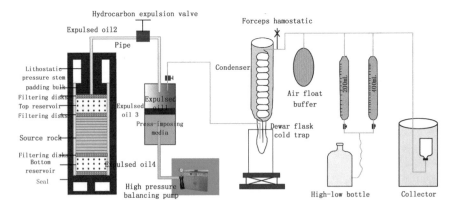

Fig. 2.21 An illustration of experimental simulation for hydrocarbon generation and explusion in pore space limited source rocks

liquid nitrogen. The oil and water mixture is frozen in the collection pipes while gases enter a displacement tube for qualification and are finally quantitatively analyzed using a gas chromatograph for composition (hydrocarbon gases and non-hydrocarbon gases).

② Collection and quantification of expelled oil

According to different sample installation patterns, expelled oil is collected from different portions of the sample. After the experiment, the first collection is a pressurized expelled oil and water mixture (expelled oil 1) and gas. When the system cools to 150 °C, the exit valve is opened to allow hydrocarbon expulsion and then gas condensate, gas and water after they travel through the gas and liquid separation and collection device cooled by ice water (oil 2). Finally, when the whole system has cooled to room temperature, the high pressure reaction container is opened, the container inside the sample chamber, exterior and inner connecting pipes are rinsed with chloroform to obtain light oil—the rest of expelled oil 2. The top and bottom sands of the sample chamber are removed and expelled oil 3 and 4 are recovered from them.

③ Collection and quantification of residual oil and solid: Bitumen (the residual oil) from the residual source rock sample is recovered after the experimental simulation. Residual oil and expelled oil comprise the total oil. Total oil and gas comprise the total hydrocarbon.

iii. Experimental results

(1) Characteristics of hydrocarbon generation and kerogen evolution in pore space limited source rocks

Figure 2.22 shows the hydrocarbon generation quantity and productivity curve of two source rock samples (Wang 24 and Pu 1-154) of different TOC content and kerogen type, in the same boundary conditions, in a totally tight environment and without hydrocarbon expulsion. Hydrocarbon generation curves under a continuous subsidence situation in pore space limited source rocks differ significantly from traditional hydrocarbon generation curves. Their remarkable characteristics are

① No 'big hump' exists as generally noticed in the traditional hydrocarbon generation curve as Ro evolves from 0.7 to 1.3 %. TOC content of Wang-24 samples differs more than three times from that in well Pu 1-154. Kerogen types of the two wells are type I and II_1, respectively. However, the samples enter a stable state of hydrocarbon generation and show no increase in productivity when simulated temperature reaches 360 °C and Ro is 0.9 %. The stable curve is maintained up to 385 °C and Ro 1.4 %. These results not only challenge traditional concepts that liquid hydrocarbon transforms to gas within the oil window, but also suggest the long duration of the hydrocarbon generation phase and consequently favors liquid hydrocarbon preservation in the deep subsurface. This can explain the occurrence of large volumes of oil but poor in gas in eastern Chinese petroliferous basins because the major source rocks of the Cretaceous and Paleogene are still in the maturity range of 0.9–1.4 %Ro.

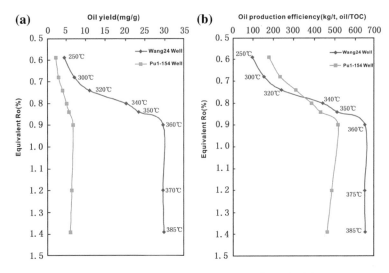

Fig. 2.22 Experimental hydrocarbon generation quantity and productivity curves to simulate hydrocarbon generation in pore space limited source rocks in continuous subsidence situation

② There is no specific low maturity stage in kerogen evolution. When programmed temperature increases from 300 to 360 °C with a small Ro increment of 0.22 % (from 0.68 to 0.90 %), hydrocarbon generation amounts and productivity of the two samples from Wang-24 well and Pu 1-154 well soar from the minimum value to the maximum. This 'giant leap' in the hydrocarbon generation process indicates the oil window can be complete over just a few hundred meters of burial depth after the threshold.

(2) Changes in fluid pressure in pore space limited source rocks

Figure 2.23 shows the variation of fluid pressure curves of two source rock sample suites with different TOC content to simulate hydrocarbon generation in pore space limited source rocks in a continuous subsidence situation. The controlling factors are high temperature and hydrocarbon generation. Pu 1-154 sample has a low organic content (TOC = 1.3 %), poor organic type (type II_2) and small hydrocarbon generation ability. During hydrocarbon generation under high pressure in a sealed environment, the Pu 1-154 sample shows no substantial increase in fluid pressure and pressure coefficient, which are 14.0–18.5 MPa and 1.07–1.48, respectively (high pressure coefficient may be fluid expansion caused by the temperature rise). The amount of hydrocarbon generation is in the range of 2.32–6.72 mg/g during the whole process. This suggests that the limited pore spaces are not completely filled with hydrocarbon in continuous subsidence situation. The generated hydrocarbons may dissolve in liquid water under high temperature and pressure conditions. Thus, no large scale pressure increment is created by kerogen conversion to hydrocarbons.

The hydrocarbon generation amount ranges from 4.45 to 29.78 mg oil/g rock in Wang-24, which greatly exceeds that of the Pu 1-154 sample when they are at the

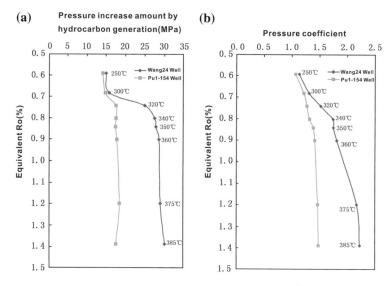

Fig. 2.23 Experimental results of pressure increment after hydrocarbon generation in source rocks

same maturity level. When the simulation temperature is 320 °C, equivalent Ro is 0.74 %, the hydrocarbon generation amount of Wang-24 black shale is 10.95 mg oil/g rock. There are obvious increases in the fluid pressure and pressure coefficient, which indicates that the pore spaces are rapidly filled with hydrocarbons, resulting in rapid increase in pressure. Sharp changes occur between 300–320 °C and Ro between 0.68–0.74 %.

Change in fluid pressure in pore space limited source rocks is a manifestation of hydrocarbon generation and variations in oil and water phase states. When generated hydrocarbon is a certain quantity or oil saturation in source rocks pore space reaches a certain value, abnormally high pressure occurs in continuous subsidence of source rocks. Based on this, hydrocarbon generation amount and the consequent source rock hydrocarbon saturation can be determined by studying changes in the paleo-pressure field, source rock pore space and paleo-temperature for a certain burial depth during continuous subsidence.

3. Computation principles and methods of hydrocarbon generation amount in pore space limited source rocks

i. Computation principles of hydrocarbon generation amount in pore space limited source rocks

Source rocks are generally formed under highly reducing depositional environments, where muddy sediment diagenesis coexists with organic matter thermal evolution. As subsidence proceeds, muddy sediments consolidate gradually, and hydrocarbon generation by organic matter thermal degradation is concomitant. The generated oil will be preserved in source rock pore space which is limited. Theoretical 'accommodation' for hydrocarbon in source rock is the whole pore space at the last stage of a basin continuous subsidence stage (before the stage of overall uplift and erosion). Thus, quantitative study of petroleum resources is actually a dynamic and systematic study of the petroleum geological evolution process of petroliferous basins. By determining source rock pore space and oil saturation at different evolutionary stages, generated hydrocarbon amounts can be calculated accordingly.

ii. Computation formula for hydrocarbon generation in limited pore space

According to basic principles of hydrocarbon generation amount in pore space limited source rocks, the quantitative formula of hydrocarbon generation is

$$Q_{generation} = \int H \times S \times F_{generation} \times S_o \times \rho$$

where

$Q_{generation}$ refers to the amount of hydrocarbon generated, t;
H refers to the effective source rock thickness, m;
S refers to the effective source rock area, km^2;
$\Phi_{generation}$ refers to the effective source rock porosity before hydrocarbon expulsion (i.e., pore space), %;
S_o refers to the effective source rock oil saturation before hydrocarbon expulsion, %;
ρ refers to the crude oil density, g/cm^3.

In the above formula, H and S can be obtained directly from oil field practical data. The parameter $\Phi_{generation}$ can be obtained from the source rock porosity-depth curve. It is necessary to explain that the determination of source rock (mudstone) top and bottom interface burial depths require not only the logging data for stratigraphic layering, but also seismic profile correlation of wells with no logging data after decompaction and the original formation thickness before compaction and hydrocarbon expulsion is recovered. Considering the lack of source rock porosity data, a great number of core tests are required.

Determination of S_o is difficult. One of the reasons is that current experimental simulations cannot generate reliable data. Additionally, few wells were drilled in source rock developed areas in previous prospecting practices, resulting in a lack of oil saturation data. Even if there are oil saturation data, they are about residual oil saturation after hydrocarbon expulsion. This involves a problem concerning the difference between original oil saturation before hydrocarbon expulsion and residual oil saturation after expulsion. Theoretically, whether these two saturations are the same or different largely depends on phase states of oil, gas and water in source rock pore space. Current research achievements tell us that oil and gas can be dissolved in water. Even if source rock is eroded in the overall uplift phase of basins, the eroded amount is limited. In addition, source rock pores remain under high pressure as oil, gas, and water are expelled as a miscible phase and compaction effects occur. Between 0.9 and 1.4 %, the hydrocarbon generation amount and productivity remain stable, and the transformation from liquid hydrocarbon to gas hydrocarbon is restricted. Therefore, the primary migration of petroleum is in the dissolved phase in water. When they enter conduits to start secondary migration or are in traps, free petroleum is formed from gradual exsolution and the phases are separated from each other. Gas travels fastest, oil is the second, and water is the last.

In light of the above findings, oil saturation data of well logging in source rock developed areas can be applied directly. To avoid error caused by oil saturation heterogeneity, it is preferred to apply an average value of data from several wells or layers. For example, average oil saturation of the 24 m source rock in the upper portion of the ES_4 at well L1 in Dongying Sag is 30 %; average oil saturation of the 20 m source rock of Hetaoyuan Formation at well Mi-1 in Dongying Sag is 28 %; average oil saturation of eight wells in Xingou Formation source rocks at Xingou area is 29 %.

2.2.4 Fundamental Issues Remaining to Be Solved

If the idea of kerogen thermal evolution for hydrocarbon generation is to take organic matter (kerogen) from petroleum geological environments into the laboratory using programmed temperature to simulate hydrocarbon generation in a Rock-Eval device and by means of geochemistry, then the way of thinking of hydrocarbon generation in pore space limited source rocks is to take source rocks the opposite way and dealing with simulated geological conditions to study hydrocarbon generation and its characteristics in the subsurface, by an comprehensive analysis of data obtained from models.

To study hydrocarbon generation and its characteristics in subsurface source rocks under geological conditions, the authors and affiliated research groups, coupled with scientific workers from Shengli, Zhongyuan, and Henan oil fields, have spent more than ten years to study geological evolution characteristics of Dongying Sag, Dongpu Depression, Miyang Depression, and Baichagan Sag. We have performed considerable fundamental researches on source rock sedimentation and its influence on porosity. Meanwhile, according to natural conditions where source rocks evolve, we have independently developed models to conduct experimental simulations using source rock samples from the mentioned sags. To study mutual dissolved states of oil, gas, and water in pore space limited source rocks, we have independently developed a method to measure multiphase fluid solubility. Based on previous research and experimental simulations, a preliminary theory on hydrocarbon generation in pore space limited source rocks is emerging.

Although the preliminary results are encouraging, there still large gaps in the ways of thinking, simulation experiments, and application. Here is a brief introduction to the relevant issues.

1. Source rock evolution history and diagenesis require more investigation

Since the end of the 1970s, source rock research focus shifted from comprehensive study of petroleum geology to petroleum geochemistry. A number of petroleum geologists working on source rocks became petroleum geochemists. Meanwhile, the theory of kerogen evolution for hydrocarbon formation has become guidance for source rock studies. Source rock research in this period gradually moved from geological conditions to laboratory studies guided by ways of geochemical thinking. The main topic and basis of assessment researches on source rocks are analytical indexes as well as source rock thickness, rock types and distributions. Excessive emphasis on geochemical indicators caused deviation of research from geological settings. This is inadequate as organic matter is just a portion of source rocks but the whole rock system.

To avoid these problems, further research on source rocks, the source rock kitchen, are required. Details are as follows:

① A further probe into deposition and diagenetic evolution history of major source rocks in the main petroliferous basins of eastern China, including sedimentary environments of source rocks, water salinity, paleontology, mineral compositions, deposition rate, rock assemblages on source rock lithological column and source rock stratigraphic and thickness isopach map, thermal history of depositional center of source rocks and source rock diagenesis evolution, in an effort to make a comprehensive analysis of source rock petroleum geochemical evolution history.

② Taking the experimental subjects of representative source rocks in Qijia-Guloong and Sanzhao sags of the Songliao Basin, Dongying and Chezhan sags of the Jiyang Depression, Miyang and Nanyang depressions of the Nanxiang Basin to systematically investigate the similarity and difference among them and to comprehensively understand petroleum geochemical evolution principles of the source kitchen in different petroliferous basins.

③ Combining with practical exploration data of major petroliferous basins in eastern China to conduct comprehensive studies on petroleum geological evolution and organic thermal evolution in an attempt to enrich and improve calibration and election of essential parameters in computation of hydrocarbon generation amount in pore space limited source rocks.

2. Research on porosity evolution in source rock kitchens

Source rock porosity evolution is the basic topic of this theory. However, as production wells were generally drilled on structural high points such as uplifts and traditional slopes in petroleum exploration practices, and favorable traps or reservoir are rare in source rock kitchens, few production wells were drilled into source rock kitchens. Source rock kitchens are still unexplored areas in petroleum exploration. Data about source rock occurrence and thickness changes are obtained by means of seismic reflection profiles, while data about source rock geochemical indexes are obtained via tests of source rocks drilled in structural traps. This reality introduces difficulties in studies on source rock porosity change, especially when borehole and core data are absent.

In recent years, along with the rapid developments in shale oil and gas explorations, there has been a considerable increase in drilling activity in source rock kitchens, which provides extensive data for studies on source rock characteristics during thermal evolution in source rock kitchens, as well as favors for further breakthroughs in this theory. The following studies are conducted basically on source rock porosity change

① Based on practical well data from oil fields in east and central China, major source rock burial depths and porosity profiles have been established. Porosity change characteristics of different source rock kitchens and influencing factors are analyzed.

② Source rock kitchens were selected with detailed data to analyze influences of geo-temperature field characteristics, subsidence rate, depositional rate, clay mineral compositions, organic matter content, and maturity evolution on source rock porosity.

③ Relationships were established between porosity and source rock diagenesis and organic matter thermal evolution. Three stages need to be considered, i.e., characteristics of porosity change in the early phase of source rock diagenesis at Ro < 0.5 %; characteristics of porosity change when source rocks get mature and generate a large amount of hydrocarbon at Ro 0.5–0.9 %; characteristics of porosity change in a stable phase of hydrocarbon generation at Ro 0.9–1.4 %. Involved in the above analysis are clay mineral evolution of source rocks, phases of preservation pore water in source rocks, and evolution characteristics of organic and inorganic pores.

3. Phase of oil, gas, and water preserved in source rock pores

The main issue of the phase of oil, gas and water in source rock pores is mutual solubility when they coexist in source rock pores under high temperature and pressure in geological conditions. It is fundamental and significant for the determination quantity of oil and gas generated in source rock pores—the container— suppose source rock pore is the container for organic chemical reactions.

To work on this topic, the authors and affiliated research groups have independently developed a method to measure multiphase fluid solubility. This device is capable of testing multiphase fluid solubility of oil, gas, and water under geological conditions by sampling at temperature pressure equilibrium. The reason to develop this device is to find out whether oil, gas, and water are in miscible states in source rock pore space by laboratory simulation. Although the results delivered did not match our expectation due to constant temperature and pressure conditions, oil and gas solubility in water are obtained. This is just one test of physical properties and no clear pattern can be observed. Once temperature and pressure adjustment was settled, more data will be generated. Although no breakthrough has been made through this experiment, miscibility of oil, gas, and water can be demonstrated in source rock media. Without this experiment, the only way to understand solubility behavior is physical and chemical methods rather than a petroleum geological approach.

To scientifically simulate phase behavior of oil, gas, and water in source rocks, the following basic research needs to be conducted.

① Develop new models for experiment in which hydrocarbon is generated in source rock pores under high temperature and pressure and oil, water, and gas are miscible. This device should be applied for simulations not only of hydrocarbon generation in source rock pores under in situ conditions, but also of the equilibrium process of oil, gas, and water.

② Select source rock kitchens with relatively enriched drilling data of shale oil or gas for further studies on diagenesis, porosity change and organic matter thermal evolution. In the application of newly developed modeling devices, simulate process of hydrocarbon generation from source rocks and miscible states of oil, gas, and water in source rock pores, and finally, determine oil saturation, water saturation and compare them to data from drilled wells.

③ Select petroliferous sags with abundant petroleum resources and enriched data for compiling oil saturation ichnography of major oil reservoirs. Select several typical locations cross source rock kitchens and petroleum reservoirs to make cross sections of oil saturation and to determine petroleum migration directions and hydrocarbon expulsion quantities.

Chapter 3
Hydrocarbon Generation in and Expulsion from Pore Space Limited Source Rocks

Hydrocarbon generation, migration, accumulation, and destruction in petroliferous basins are the consequences of multiple physical and chemical reactions between sediments and organic matter during geological evolution. These dynamics are ongoing in the geological settings of basin subsidence, overall uplift and shrinkage. The basin subsidence is perceived as sediment buildup, energy transformation and accumulation, i.e., hydrocarbon generation processes, featured by a physical field of sediment load and increasing pressure, which has been discussed in chapter two. The overall uplift stage of a basin is perceived as the release of accumulated energy, i.e., hydrocarbon accumulation process with a physical field of sediment unloading and pressure release. The shrinkage stage of a basin is perceived as energy adjustment and equilibrium of material compensation, i.e., completion of hydrocarbon accumulation with a physical field in equilibrium.

This chapter focuses on a discussion of the variations in the physical field of source rock occurrence in the depocenter and reservoir occurrence in basin slopes during basin overall uplift stage, highlighting the decisive role of pressure and differential pressure in oil and gas expulsion and accumulation. In addition, the application of the quantification of thickness of eroded sandstone by means of astronomical Milankovitch Cyclicity and the important concept "Sandstone Relaxation" which is closely associated with hydrocarbon accumulation are introduced systematically in this chapter.

3.1 Hydrocarbon Generation in and Expulsion from Pore Space Limited Source Rocks

This section is mainly concerned with the theoretical considerations, experimental simulations, and quantitative methods of hydrocarbon expulsion from pore space limited source rocks.

© Petroleum Industry Press and Springer Science+Business Media Singapore 2017   69
D. Guan et al., *Theory and Practice of Hydrocarbon Generation within Space-Limited Source Rocks*, Springer Geology, DOI 10.1007/978-981-10-2407-8_3

3.1.1 Theoretical Consideration of Hydrocarbon Expulsion from Source Rocks

The motion and development process in the universe is from imbalance to balance and to imbalance again. Imbalance is eternal and absolute while balance is short and relative. This principle is applicable in petroliferous basin. After the subsidence stage, a great mass of materials and energy are accumulated. The constant aggregation will definitely cause imbalance between basin center and marginal areas, to which the basin will undoubtedly respond by overall uplifting and erosion to regain the state of equilibrium. The characteristic of the basin overall uplift stage is material adjustment by denudation, which makes the physical field of the basin transform from loading and pressure growth to unloading and pressure release. In the process of mass unloading, oil, gas, and water readjust to seek new balances. From the perspective of petroleum geology, the process of seeking balance is the processes of hydrocarbon generation, migration, and accumulation.

1. Characteristics of physical field at the basin overall uplift stage.

The petroliferous basin overall uplift stage is implied by denudation of strata, which intensifies in sandstone areas with the maximum occurrence in the basin margins, and the minimum or even no denudation occurring in source rock developed region. The main reason for this phenomenon is different physical properties at different areas in a basin.

(i) Characteristics of physical field at sandstone (reservoir) developed area.

With denudation during the basin overall uplift stage, overburden strata of the sandstone developed area was eroded gradually with sandstone decreasing in thickness, rendering this area into the unloading and decompaction physical field. Consequently, the massive elastic potential energy within the sandstone will be released in the direction of unloading (upward direction). In this case, two upward forces exist in the sandstone developed area. One is the driven force of basin uplift and the other is inner force of energy accumulated within the sandstone which is called the sandstone relaxation force. With the co-effecting of the two forces, the uplift of the sandstone developed area is faster, so the quantity and speed of denudation are larger, and the lost strata are more. With the continuous unloading process by denudation at the basin overall uplift stage, an internal continuous decompaction environment forms in the sandstone developed area. So during the denudation in the basin overall uplift stage, under the control of unloading and decompaction physical field in the sandstone developed area, a relatively lower pressure area was formed (Fig. 3.1).

(ii) Characteristics of physical field in the source rock (mudstone) developed area.

As the basin was uplifted as a whole, hydrocarbons of high pressure within source rocks were induced by momentary energy release to expand under a decreased

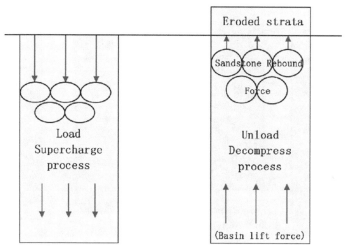

Fig. 3.1 Characteristics of physical field in sandstone developed area during the basin overall uplift and erosion stage

pressure regime, and consequently migrated away from the source rock when fractures occur. Oil and gas from internal source rocks migrate to low pressured sandstones. Sandstones interbedded in mudstone or faults linking source rock and sandstone became the conduit for petroleum migration. Because of a fluid loss within source rock pores, source rocks were compacted further and have their porosity decreased by accumulating overlapping strata. Thus, there exist two sources of force in source rock developed area: one is the driven force of basin elevation, which is upward and the other is downward from the overlapping sediments which constantly aggregated because of fluid migrating away from the source rocks. Source rocks will be further compacted with decreased thickness. Under the co-effecting of these two forces, the source rock developed area has not only the smallest rate of elevation, but also the minimum denudation in the basin, and tends to subside to receive sediments rather than rise to get eroded. In conclusion, except for an unloading and decompaction state in the early period of this stage in the basin overall uplift stage, the source rock developed area features uploading and pressure increment and forms areas with relatively high pressure (Fig. 3.2).

In summary, during the basin overall uplift and denudation stage, sandstone developed areas form a low pressure zone due to sediment unloading and decompaction, while source rock developed areas keep high pressure due to loading. The physical field during the petroliferous basin overall uplift stage provides the driving force for oil and gas primary migration. The differential pressure between source rocks with high pressure and reservoirs with relatively low pressure drives oil and gas to migrate.

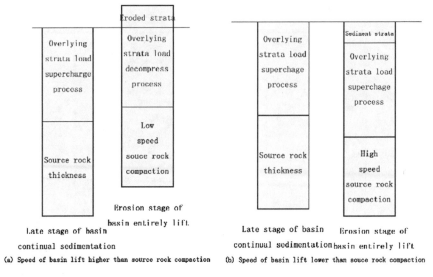

Fig. 3.2 Characteristics of physical field at source rock (mudstone) developed area in overall uplift and erosion stage of a basin

2. Pressure regime at the basin overall uplift stage.

Basin transformation from subsidence to overall uplift and denudation is a crustal material equilibrium process between the basin and surrounding areas. This process can be regarded as buildup, adjustment, and ultimately relative balance of materials (including solid, liquid, and gaseous) in the form of hydrocarbon accumulation. Because of the strong mobility of oil and gas, their migration speed and direction are highly constrained by regional pressure regimes, from the higher pressure area to the lower pressure area. When regional pressure fields reach balance, oil and gas will remain relatively immobile. As a result, the essence of the subject of hydrocarbon accumulation is basin pressure regimes, especially in a source rock area and a reservoir area during the basin overall uplift and denudation stage.

(i) Analysis of pressure regimes in source rock (mudstone) developed area.

The major source rocks in the source rock developed area have already entered the oil window during the last phase of basin subsidence. With the compaction effect of overburden rocks, pores filled with oil and gas were in the state of abnormally high pressure. When the basin was transformed from subsidence to overall uplift and denudation, under the effects of the unloading and decompressing physical field, source rocks were fractured instantly by ultrahigh pressure hydrocarbons filling the pores, which drive hydrocarbons to migrate out from the source rocks. Hydrocarbon migrating through tiny fractures of the source rock has been named microfracture expulsion. As pore fluids were expelled gradually, porosity of source rocks decreased as compaction intensified. This constant compaction of source rock caused by fluid expulsion was called 'compaction caused by hydrocarbon expulsion.' Source rocks enter a stable stage of 'expulsion driven by compaction' under the compaction of overburden rocks. Two special cases need to be addressed in 'expulsion driven by compaction.'

① The speed of the basin overall uplift and denudation larger than source rock compaction rate.

In this case, source rock overburden strata were denuded gradually and got thinner, relieving the compaction of source rocks, and consequently decreasing the rate of 'expulsion driven by compaction,' as well as weakening the driving force for hydrocarbon migration (Fig. 3.3a). This condition is not favorable for hydrocarbon expulsion from source rocks or migration.

② The speed of the basin overall uplift and denudation smaller than that of source rock compaction.

This case is caused by fast expulsion of oil and gas from source rocks, which leads to decreasing source rock porosity and rapid compaction. Hydrocarbon expulsion creates an instant pressure reduction in the source rock pores leading to a slow rate of subsidence of regional source rocks as compensation for this lower pressure. However, the overburden thickness continues to increase and compaction intensifies, which in turn largely increases the rate of hydrocarbon expulsion driven by compaction (Fig. 3.3b). This probably accounts for the small eroded thickness and good preservation of source rocks in the basin overall uplift and denudation stage.

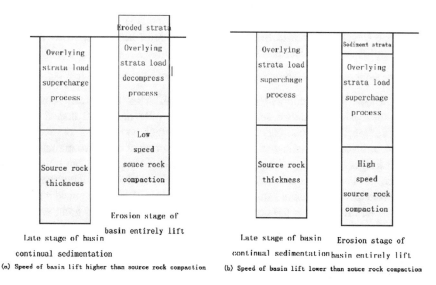

Fig. 3.3 An illustration of 'hydrocarbon expulsion driven by compaction' from source rocks

In summary, microfracture expulsion and compaction caused by hydrocarbon expulsion take place in a source rock in the basin unloading and decompaction physical field during the basin overall uplift and denudation stage. When source rocks enter the stable stage of compaction caused by expulsion, effective source rocks preserved high pressure under the gravity compaction of over burden rocks, and basin has a high pressure area. If several source rock sequences were developed in a basin, all of them are situated in high pressure regions. As long differential pressure between these regions and the sandstone developed areas are maintained, hydrocarbons will migrate to the low pressure areas and accumulate constantly.

(ii)   Analysis of the pressure regime in the reservoir (sandstone) developed area.

The reservoir developed areas in continental sedimentation basins are mainly controlled by sandstone spatial distributions in various fluvial delta and fan delta areas. Due to differences in sandstone thickness, grain size, roundness, sorting, cement compositions and types, accumulated elastic potential energy varies significantly in the basin subsidence stage. When the basin transformed from subsidence to uplift and began eroding under the effects of the unloading and decompaction physical field, the speed of sandstone relaxation upward (elastic potential energy releasing upward toward earth surface) differs with locations, which causes differences in the uplifting process in different places in the sandstone developed area. If subjected to multiple external forces during uplift and denudation, the reservoir spatial distribution and geography could change dramatically, forming different geological structures and trap types. These geology structures or traps were formed by sandstone and interbedded mudstone. As long as differential pressure existed between these geological units and source rocks during the process of basin overall uplift and denudation stage, oil and gas will certainly migrate into these geological structures and traps. In conclusion, the pressure regime of the sandstone developed area is the key for hydrocarbon accumulation in the basin overall uplift and denudation stage.

(1)   The concept of sandstone relaxation and its petroleum geological significance.

As we know, sandy sediment is elastic material. After mechanical compaction during the basin subsidence stage, sandy grains were tight, and turned into the elastic compaction development stage under further compaction of overburden rocks. A basic characteristic was the gravity compaction of overburden rocks so that they no longer behaved in the manner of mechanical compaction but preserved in the form of elastic potential energy of sandy grains. As long as the external force has not reached the pressure limit of sandstone rapture, the internal elastic potential energy would increase as the compaction continued. Therefore, a great amount of elastic potential energy had been accumulated within sandstones under gravity compaction of overburden rocks during the last stage of basin subsidence. Obviously, sandstone which had a large proportion of quartz, large thickness, good roundness, good sorting, and cementation would have high accumulated elastic potential energy.

The huge elastic potential energy accumulated within sandstones will be released under the effects of unloading and decompressing when the basin transformed from subsidence to overall uplift and denudation. The potential released toward earth surface was called sandstone relaxation. This phenomenon occurs not only in continental sedimentary basins but also frequently in marine clastic sedimentary basins. The reasons why sandstone relaxation is important in oil and gas accumulation process are as follows:

① The sandstone relaxation is energy released from sandstone during the basin uplift stage, which creates a low or relatively negative pressure region in the sandstone developed area, resulting in a target destination of mature oil and gas migration. The tremendous pumping power exerted on source rocks acts as the basic driving force triggering hydrocarbon secondary migration.

As described in the previous section, source rocks were broken and oil and gas were released at the initial stage of uplift and denudation due to unloading and decompressing, resulting in more compaction. The source rocks entered the stable stage of expulsion caused by compaction under the overburden rocks. The source rock developed area may partially stay in the subsidence stage in spite of overall uplift. Pressure exerted on the effective source rocks drives oil and gas expulsion, similar to a syringe piston (Fig. 3.4). The sandstone area functions in the totally opposite way. The sandstone developed area works as drainage area during the basin overall uplift and denudation stage due to unloading (erosion) and decompaction (pressure release). Sandstones are the favorable direction of oil and gas migration discharged

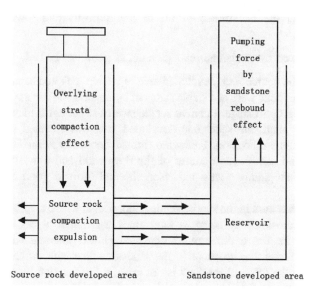

Fig. 3.4 Characteristics of the pressure field in a source rock area and sandstone area during the basin overall uplift and denudation stage

from overpressured source rocks. Moreover, sandstone relaxation is a process of decompression in the form of elastic potential energy release. The upward resilience force was a kind of suction similar to a syringe piston withdrawal. Under the effects of these two factors, oil and gas migrates from the source rock area to the sandstone area and consequently accumulates.

② Sandstone relaxation improves the quality of sandstone reservoirs.

The sandstone reservoir quality assessment generally includes the analysis of depositional conditions and facies, system characterization, digenetic stage classi-fication, and primary pore and secondary pore formation mechanisms. This method is based on the perspective of sedimentology rather than petroleum geology as basin evolution history is not included. The core samples and reservoir are products of basin evolution in the overall geological history rather than a certain evolution stage. Currently, hydrocarbon accumulation mainly at the stage of basin overall uplift and denudation stage is largely accepted; therefore, assessment of sandstone reservoir quality should be based on the characteristics of the physical field in this evolution stage.

During the process of basin overall uplift and denudation stage, sandstone relaxation played an important role in improving the quality of the reservoir by increasing porosity and permeability, manifested by changing grain contacts from tight to loose and the loss of cement. The loosening grains of sandstones and loss of cement were observed from reservoir cores in Daqing and Shengli oilfields. Thus sandstone reservoirs would be analyzed and evaluated realistically only in light of the physical field characteristics of basin overall uplift and denudation stage. The reservoir quality at this period was better than the current situation; otherwise, it is hard to explain the occurrence of oil in low porosity and low permeability reservoirs.

③ Faults formed by sandstone relaxation act as good carrier bed.

In the case of the source rock developed area or source rock and sandstone contact zone, the source rocks were at a stable status of compaction driven expulsion during the basin overall uplift stage. Relative subsidence or slight uplift may occur in this area. On other hand, the sandstone developed area was elevated at high speed owing to sandstone relaxation. Faults are formed frequently at the contact area of source rocks and sandstones because of the downward and upward forces. This kind of fault links source rocks and reservoirs and forms a good carrier system (Fig. 3.5).

Another case occurs in the sandstone developed area, where a variable degree of relaxation and erosion may occur as sandstones at different places have different thickness, and in turn a different accumulated elastic potential energy. In this physical field, faults tend to occur in the sandstone area with different relaxation rates. Although the offset of the fault is not large, it forms a link between different sandstone layers, and thus acts as crucial secondary migration carrier.

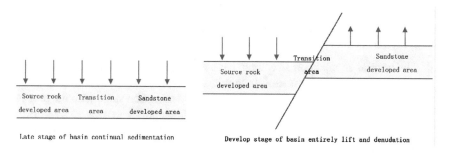

Fig. 3.5 An illustration of a fault linking source rocks and sandstones in the basin overall uplift and denudation stage

(2) Differential compaction structures and hydrocarbon accumulation plays.

Differential compaction structure is a sedimentary phenomenon formed during basin subsidence. Various fluvial deltas and fan deltas reached to the lacustrine basin center, resulting in a sandy rich belt surrounded by muddy sediments. Differential compaction structures formed because of the discrepancy between mechanical compaction and elastic compaction. The term differential compaction structure was created to highlight its genetic mechanism. The sandy sediments of differential compaction structures are composed mainly of quartz and feldspar grains with good sorting, high roundness, contact cementation, and low cement contents. Those sandy sediments would become high quality reservoirs after diagenesis.

During the basin subsidence stage, the coast line moves back and forth due to fluctuation of lake level, which in turn would cause vertical variation in sandstone distributions. If a river changes its direction, the lateral distribution of channel deposits would also change. So in the range of sandy sediments which were controlled by a certain sedimentary system development stage, differential compaction structures only develop in the area with maximum superposition thickness and best quality sandy sediments. If an independent sediment system has several areas with favorable conditions for developing differential compaction structures, they may occur in local peaks of individual sandstone rich areas.

Thus, differential compaction structures actually are an indicator of sandstone distribution and maximum thickness of sandstones in sandstone developed areas. Their occurrence is a sedimentary phenomenon and does not make sense to structural geology. However, since an isopach map of strata thickness is applied to illustrate paleo-structure in certain geological time in the practice of structural geology, isopach map of sandstone thickness of certain strata in certain geological time is applied to describe differential compaction structures.

At the last stage of basin subsidence, the area of differential compaction structures not only had the maximum thickness of sandstone and best quality reservoir, but also had the maximum accumulated elastic potential energy. Thus, the effect of sandstone relaxation takes place intensely in differential compaction areas in response to the unloading and decompaction physical field when the basin entered the uplift and denudation stage.

① Differential compaction structures have the maximum elastic potential energy accumulated, the largest sandstone relaxation resilience and the largest elevation speed during the basin overall uplift and denudation stage with the process of unloading and decompression. Vertical fluctuations generated by various speeds of sandstone relaxation may form a complete anticline structure in sandstone layers. Tilted strata and faults would be developed in differential compaction structure area, presented in different attitudes and topography. Thus, various types of traps would come into being in the basin overall uplift stage.

② Differential compaction structures release maximum elastic potential energy during sandstone relaxation, and have a maximum degree of decompression, and thus have the maximum differential pressure from source rock areas, and the largest suction force generated by sandstone relaxation. These factors lead to differential compaction structures being the main target for oil and gas migration, and make several types of oil and gas accumulations surrounding the differential compaction structures. The structures themselves could act as anticlinal traps and the nearby sandstones could act as various stratigraphic and lithological subtle traps based on their diverse occurrences.

According to the above mentioned analysis, major decompression occurs in sandstone areas and the maximum decompression occurs in the differential compaction structure area. Various types of hydrocarbon accumulations surround the differential compaction structures, and this decompression system is called a hydrocarbon accumulation play. If several differential compaction structures developed in the basin, then several hydrocarbon accumulation plays may exist. Each play has different degree of differential compaction and distance from major source rocks. Different differential pressure consequently results in different hydrocarbon richness. Apparently, the larger the area of differential compaction structures and the thickness of the accumulated sandstones, the larger is the amount of oil and gas accumulated, such as Lamadian, Sa'ertu, and Xingshugang oilfields in Daqing. All of them are large oilfields formed in the geological setting of differential compaction structures (Fig. 3.6). Thus, several differential compaction structures formed during the basin subsidence could be good reservoirs for oil and gas in the accumulation process during the basin overall uplift and denudation stage. Differential compaction provides internal energy for hydrocarbon migration into various types of traps especially anticlinal traps. Following this theoretical way of thinking, differential compaction structures were not only a kind of sedimentary structure which formed by enriched sandstones, but also an independent decompression system during the process of unloading and decompression in basin overall uplift and denudation stage. This is an independent play formed during the oil and gas accumulation process. No matter where the source rocks are situated and what

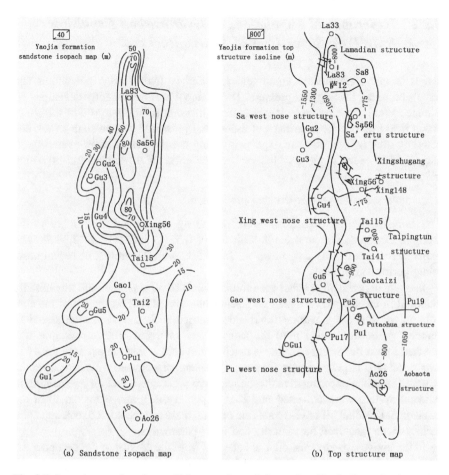

(a) Sandstone isopach map   (b) Top structure map

Fig. 3.6 Isopach map of sandstone thickness and morphology of top Yanjia Formation in Daqing (after Hengjian Wang)

kind of migration routes are used, hydrocarbon accumulations belong to one play as long as they were controlled by one decompression system of differential compaction structures.

In summary, transition from subsidence to overall uplift under unloading and decompaction physical field comprises the only requirement for hydrocarbon accumulation process in a petroliferous basin. The overall uplift and denudation stage is the main period for pool formation. As for the small amplitude of vertical movement when the basin transformed from overall uplift and denudation to shrinkage, it was only regarded as completion continuation of the hydrocarbon accumulation process. Because of the continuity of oil and gas accumulation, once the basin entered the main hydrocarbon accumulation period, this process will carry on until the Quaternary.

3.1.2 Experimental Simulations of Hydrocarbon Expulsion from Pore Space Limited Source Rocks

DK-II is our experimental equipment to simulate hydrocarbon generation and expulsion under heating and pressure. The device was independently developed by Sinopec Wuxi Petroleum Geology Research Institution. The purpose of the simulation is to find out the influences of differential pressure between source rock and reservoir area on hydrocarbon expulsion during the process of basin overall uplift stage, and to examine the correlations between differential pressure and expulsion efficiency.

1. Sample selection and experiment procedures.

The samples for the experiment were picked up from the Es3 member in the Dongpu Depression at well Hu88 with Ro of 0.52 % and TOC of 2.31 %. The experimental conditions are shown in Table 3.1, with 23 groups of experiments being performed in total.

In every experiment, a small core column made of 60 g source rock was initially enclosed in the sample cell, then 20 g sandstone was placed on the top of the core column to simulate a natural interbedded source rock and sandstone situation. Finally, water was added until the sample was saturated. The temperature was increased until the designed value is reached and held at that temperature for 48 h. Gas 1 and oil 1 expelled under differential pressure effects were collected. After the completion of the experiment at this temperature point, gas 2 and oil 2 remaining in reaction system were collected and then expelled oil 3 was extracted from the sandstone. Residual oil refers to soluble organic matter in the source rock and solid residues were analyzed for maturity and other parameters.

In the above experiments, oil 1 is collected from the expulsion devise under the differential pressure ($\Delta P = 0$, 3, 6 and 12 MPa) between generation system and reservoir system. It is similar to the natural case in which oil is generated in source rocks and migrated to effective reservoirs after certain distance. Oil 2 is oil retained on the surface or microfractures of source rocks after hydrocarbon expulsion. Oil 3

Table 3.1 Experimental conditions to simulate hydrocarbon expulsion under differential pressure in basin overall uplift stage

Simulation temperature (°C)	Simulation time (h)	Lithostatic pressure (MPa)	Formation pressure (MPa)	Differential pressure for expulsion (MPa)
250	48	48	21.5–26.2	0, 3
275	48	57	26.5–29.2	0, 3
300	48	58	27.8–31.5	0, 3, 6
320	48	64	30.6–41.5	0, 3, 6, 12
340	48	72	32.2–43.2	0, 3, 6, 12
360	48	92	38.8–50.4	0, 3, 6, 12
375	48	100	42.5–53.0	0, 3, 6, 12

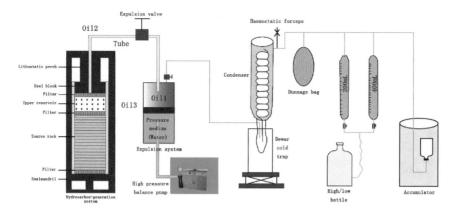

Fig. 3.7 An illustration of experimental setup to simulate hydrocarbon expulsion under differential pressure effects

is oil retained in the sandstone on top of the source rock in hydrocarbon generation and expulsion system, which is similar to oil in the interbedded sandstone layers in natural occurrence source rocks (Fig. 3.7).

2. Experimental results and their petroleum geological implications.

Oil 1 shows few variation with increasing pressure and temperature under no differential pressure (similar to the basin subsidence stage) (Fig. 3.8). Oil expulsion yield remains low with an expulsion coefficient about 1.0 % even when the oil window is reached. This suggests mature source rock alone does not necessarily mean generated hydrocarbon migration into reservoirs or traps without obvious uplift taking place in the basin or sags.

Meanwhile, when the programmed temperature was lower than 320 °C (Ro < 0.7 %), both oil 1 yield and expulsion coefficient were very low even if there was differential pressure between the source rock system and the expulsion system. This suggests that if main source rock in the basin or sag did not enter the oil window during the end of basin subsidence stage (before the overall uplift stage), as the quantity of generated oil was limited, so is oil saturation in source rock pores and amount of increased pressure by hydrocarbon generation. In this case, even if there is differential pressure between source rocks and reservoirs generated by basin uplifting, the long distance migration from source rocks to reservoirs or structural traps and accumulation would be difficult. When the programmed temperature reached 320 °C (Ro > 0.7 %), with the increase of the programmed temperature (source rock enters the oil window) and differential pressure between generation and expulsion systems, the yield and expulsion coefficient of oil 1 increased as well. In addition, results of the experiment with differential pressure of 3 MPa were similar to those of experiments with zero differential pressure, and results of

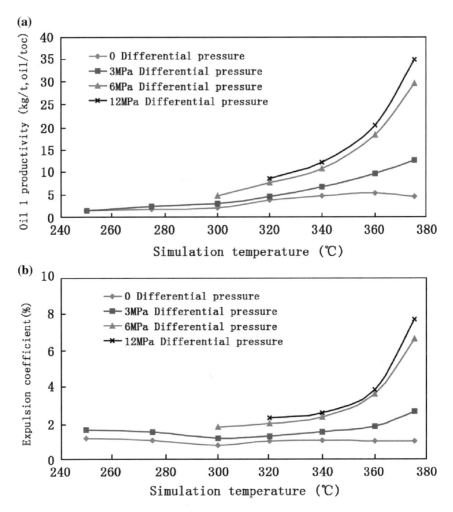

Fig. 3.8 Oil 1 yield and expulsion coefficient under different differential pressures

experiments with differential pressure of 6 MPa were similar to those of 12 MPa. This implies that when the main source rocks in a basin or sag enter oil window before the overall uplifting stage but differential pressure between the source rock area and reservoir area is limited, oil generated from source rocks could hardly migrate a long distance to the reservoir or structural traps and accumulate. When the differential pressure reached a certain value (i.e., 6 MPa), although the additional increase of differential pressure does not significantly increase yield and expulsion coefficient of oil 1, long distance oil migration from source rocks to reservoirs and accumulation occur, which means there is critical differential pressure (4–5 MPa) to drive oil migration during the basin overall uplift stage.

As Fig. 3.9 shows, the yield and expulsion coefficient of oil 2 increase with increasing programmed temperature. This tendency seems only related to temperature but differential pressure. This suggests that hydrocarbon generation has completed by the end of basin subsidence, with a large proportion of generated oil remained on surfaces and in microfractures of the source rocks. Although some strata were eroded during the overall uplift and denudation stage, oil was expelled continuously from source rocks under various differential pressures and oil 2 has no relationship with differential pressure in general.

The yield and expulsion coefficient of oil 3 were similar to those of oil 2 (Fig. 3.10), which is related merely to temperature rather than differential pressure. It implies sandstone interlayers and source rock micropores/factures were the primary reservoir space for oil and gas.

Before source rocks enter the rapid oil generation stage, oil 3 has no obvious relation with differential pressure. However, when differential pressure reached

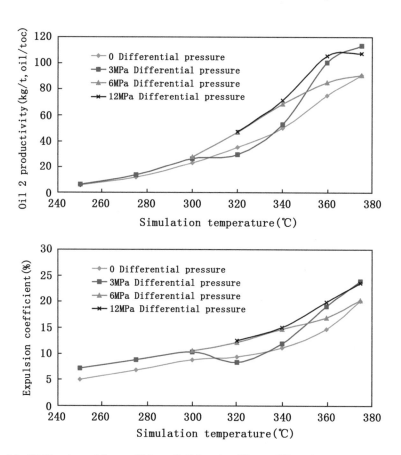

Fig. 3.9 Yield and expulsion coefficient of oil 2 under different differential pressures

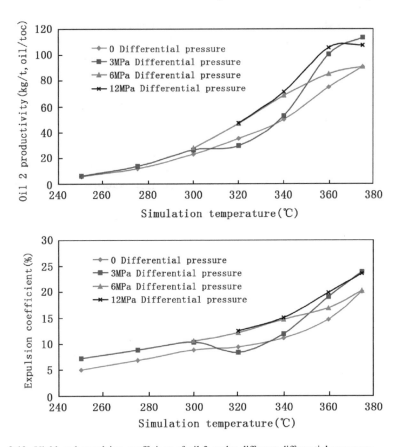

Fig. 3.10 Yield and expulsion coefficient of oil 3 under different differential pressures

6 MPa after rapid oil generation starts, yield and expulsion coefficient of oil 3 increase rapidly, which meant oil accumulation in sandstone interlayers within source rocks has an obvious relationship with differential pressure as oil 1 does.

In summary, during the basin subsidence stage, oil expelled from source rocks mainly remained on surfaces and in connected microfractures of source rocks, with part of it entering sandstone interlayers within the source rocks and possibly forming unconventional oil accumulations. Although source rocks in a basin or sag may enter the peak oil generation stage, long distance oil migration from source rocks to reservoirs and accumulation could be difficult if there was no obvious structural uplift to generate differential pressure between the source rocks and main reservoirs. Only if differential pressure between the source rock and reservoir areas reached the critical value, oil expelled from the source rock would travel a long distance to the reservoir and accumulate.

3.1.3   Principle and Method of Quantification of Oil Expelled from Pore Space Limited Source Rocks

1. The principle of quantification of oil expelled from pore space limited source rocks.

The principles of quantification of oil expelled from pore space limited source rocks can be described as follows. As burial depth increased during basin or sag subsidence, pore space of source rocks decreases gradually. At the same time oil and gas generated through organic matter thermal evolution increases, which makes oil saturation in the source rock pore space increase gradually until overpressure results and remains. The largest oil saturation in the pore space occurs when oil generation yield reaches a plateau (Fig. 3.11). During the basin overall uplift stage, internal energy within source rocks is released by basin uplifting. Consequently, various microfractures are created, through which fluids are expelled and migrate to low pressure sandstone area. Finally, source rocks are further compacted by overlying strata, causing a decrease in porosity. By calculating pore volume differences between the last period of source rock subsidence and post expulsion and compaction period, the total volume of fluid expelled from source rocks can be obtained, therefore, the total amount of oil expelled from the source rock can be calculated by multiplying oil saturation and oil density.

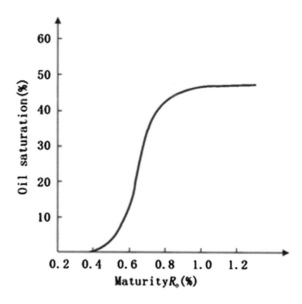

Fig. 3.11 Correlation curves between source rock maturity and oil saturation

2. Computation formula of oil quantity expelled from pore space limited source rocks.

According to the principle, model of simulation is built as follows:

$$Q_{\text{Expulsion}} = \int H \times S \times \Delta\Phi \times S_o \times \rho$$

In the formula, $Q_{\text{Expulsion}}$ refers oil expulsion amount before basin overall uplift stage;

H refers thickness of effective source rock;
S refers area of effective source rock;
$\Delta\Phi$ refers porosity difference before and after hydrocarbon expulsion from effective source rock;
S_o refers oil saturation in limited pore space during oil expulsion from effective source rock;
ρ refers density of crude oil, g/cm^3.

3. Computation procedures of oil expulsion from pore space limited source rocks.

The computation procedures are shown in Fig. 3.12, which mainly includes the following aspects:

Step 1: quantify the porosity differences between the last period of the subsidence stage and overall uplift and denudation stage or the present according to the characteristics of source rock pore evolution; Step 2: input valid values of thickness and oil saturation of selected source rock into the mathematical model for simulation. Step 3: calculate quantity of oil expulsion. Step 4: repeat the above procedures if several source rocks exist in the basin. Step 5: sum up oil expulsion

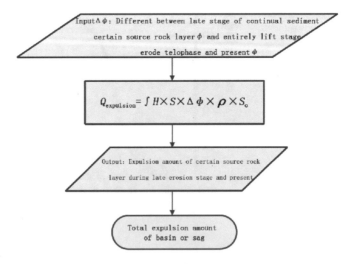

Fig. 3.12 Computation procedures of hydrocarbon expulsion from pore space limited source rocks

amounts from each series of source rock after erosion or at the present to get total amount of oil expelled in the basin or sag.

3.2 Method and Application of Astronomical Milankovitch Cyclicity in Erosion Amount Qualification

Recent research of international chronostratigraphy combines the methods and principles of the Earth's orbital (Milankovitch) parameters (precession, obliquity and eccentricity with geological chronological scale. Cyclical change of earth orbital parameters control Earth's climate change through variations in insolation (Hays et al. 1976; Imbrie et al. 1984). These changes were preserved as records in earth's sediment system. Sedimentary record includes 'paleoclimate proxies' like lithology, sediments, biology, stable isotope and well logging, through which chronology determination and correlation can be conducted with more accuracy. This approach was applied in Cenozoic stratigraphy and qualification of hydrocarbon generation and expulsion research in Shengli, Zhongyuan and Biyang oilfields and proved fruitful.

3.2.1 Basic Elements of Astronomical Milankovitch Cyclicity Theory

1. Earth's orbital (Milankovitch) parameters (precession, obliquity and eccentricity) and Milankovitch periodicity.

In 1920, based on previous achievements on astronomy and computer applications, Milankovitch from Yugoslavia proposed his hypothesis on the causes of the Quaternary ice age. Earth orbital (Milankovitch) parameters are eccentricity (e), obliquity (ε) and precession (p), which determines earth's climate pattern by periodical changing and consequently affecting insolation of places at different latitudes (Fig. 3.13). Milankovitch Cyclicity and variations of periodicity itself showed some regulating patterns, which were recorded in sediments.

Astronomers built up various orbital models of earth to simulate its periodical change. According to the astronomical periodicities since 10 Ma ago, main long eccentricity periodicities were 404.8, 95, 124, 99, 131, 667 and 1000 ka, short obliquity periodicity were 41, 39.6 and 53.6 ka, short precession periodicity were 23.66, 22.37, 18.96 and 19.10 ka (Fig. 3.14). Long period eccentricities, especially the one with maximum amplitude of 405 ka were not influenced by tide motion, and it is the most stable orbital parameter in astronomy. So eccentricity periodicity around 400 ka is the very 'pendulum' of geology time and the 'scale plate' in geochronology.

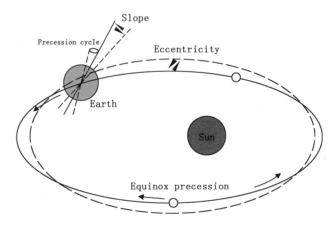

Fig. 3.13 An illustration of earth's orbital parameters (Pisias and Imbrie 1986)

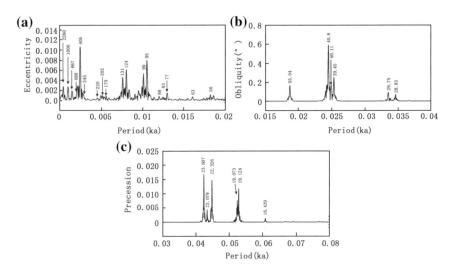

Fig. 3.14 Important orbital (Milankovitch) parameters of earth (Hinnov et al. 2005)

2. Correlations between paleoclimate proxy and astronomical periodicities.

In the middle of the 20th century, Emiliani (1955) and Hays (1976) analyzed paleontology, oxygen stable isotopes and filtering and frequency spectrum of drilling data in the southern Indian Ocean. They discovered a spectacular coincidence between the analytical data of ocean surface temperature (Ts) in summer and $\delta^{18}O$ of *Cycaldophora davisiana* 450 Ka ago and targeted curves of astronomical precession of 23.6 Ka and ecliptic obliquity of 40.8 Ka (Fig. 3.15).

Fig. 3.15 Climate variation
frequency corresponding with
ecliptic obliquity and
perihelion (Hays et al. 1976)

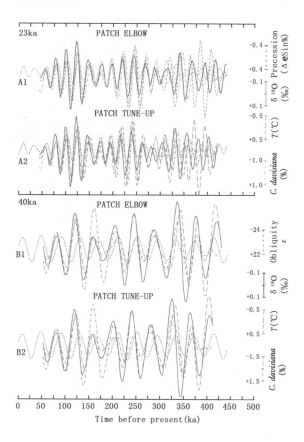

Thus, in present Milankovitch research, the common practice to apply climate proxy like variation patterns of $\delta^{18}O$, $\delta^{13}C$ and paleontology in extracting sediment cyclicity and astronomical periodicity is viable.

3. Formation and application of Milankovitch Cyclicity and magnetic polarity on the geological time scale.

Milankovitch Cyclicity is combined with high precision radiometric age and magnetic polarity strata to form an astronomical polarity geological time table (APTS) (Kent 1999). This chart is a precession and ecliptic obliquity climate change period modulated by earth's magnetic reversals measured in the seabed and orbit eccentricity (Fig. 3.16). According to this chart, a magnetostratigraphic age interval of 5332 ka, shows 130 ecliptic obliquities of 41 ka, and 13.2 eccentricity periodicities of 404 ka, and 56.1 eccentricity periodicities of 95 ka. So astronomical periodicity determined by ecliptic obliquity and precession coincided with

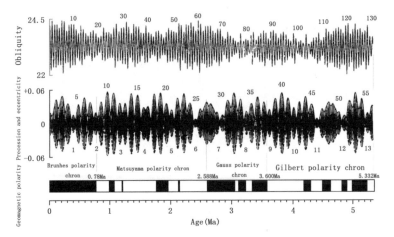

Fig. 3.16 Comparison between geomagnetic sequences, eccentricity, and precession with orbital obliquity variation periodicities (modified from Kent 1999)

geological age determined by magnetostratigraphy. This implies that the stratigraphic age can be obtained by the comparison of magnetostratigraphic time with Milankovitch Cyclicity.

Correlations of climate frequencies by numerical filtering analysis of different time scales (PATCH EIBOW and PATCH TUNE-UP modes) of ocean surface temperature in summer since 450 Ka (T), oxygen isotope (δ^{18}O), Radiolarida abundance of *C. davisiana*, procession of 23 Ka and slope of 40 Ka (B1, B2).

4. Comparison between long period timing with geological development stage.

The stability analysis of long periodicity of 405 ka is helpful in understanding earth evolution. The calculation of orbital functions provides knowledge about earth's climate change as well as references for the calibration of geological age. The resolution in the calculation of the Cenozoic by presently existing mode of long periodicity eccentricity has been controlled within 200 ka.

The division of the international standard stratigraphic time table is based on a long eccentricity periodicity of 405 ka. The younger strata geological time is counted from cardinal number 1 backward with the weakest amplitude in long eccentricity periodicity about 10 ka from the present time. The further back in time, the bigger is the periodicity number and the older is the strata. The Cenozoic contains eccentricity long periodicities 1–162 with a bottom age of (65.5 ± 0.3) Ma. According to the calculation, the Neocene has 1–62 eccentricity long periodicities and the main strata boundary at 25 Ma is clear (Fig. 3.17). Eccentricities of 100 ka and 405 ka have implications in multiple geological records especially in physical and chemical paleoclimate parameters of Deep Ocean Drilling. Therefore,

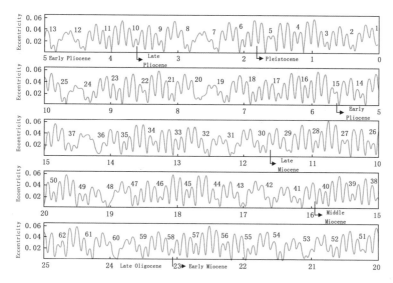

Fig. 3.17 Long periodicity eccentricities1–62 and main strata boundaries in the last 25 Ma (Lourens et al. 2005; Gradstein et al. 1995)

calculation of 405 ka long periodicity eccentricity has been applied as a geological timing scale template. For example, the paleomagnetic age of Paleocene/Neocene boundary is 23.8 Ma obtained in 1995, while according to the astronomical long periodicity of 405 ka, the point E58 with smallest amplitude has a new international stratigraphic time of 23.03 Ma determined in 2005. So new geochronological frame not only provides a more accurate comparison platform in the determination of sediments age and the analysis of geological events, but also indicates a new era of orbit stratigraphy.

Knowledge of the geological development stages from 51 Ma to present can be obtained by wavelet analysis of earth eccentricity. Yin and Han (2007) applied earth orbit eccentricity periodic change value reported by Laskar (2004), work out peak and valley of periodic fluctuation and oscillator intensity change, displaying four main periodic energy changes at 100 ka, 400 ka, 1 Ma and 2.3 Ma (Fig. 3.18). The application of long period timing scale template forms an astronomy stratigraphic table that provides important ideas of petroliferous basin subsidence, uplift and shrinkage stages, astronomy stratigraphic time and strata erosion amount calculation. For example, according to wavelet analysis, age at the lowest amplitude E80 of astronomy long period of 400 ka is about 32 Ma and age at the lowest amplitude E84 of 100 ka periodic energy change is 33.6 Ma. These two ages reflect the bottom of the Dongying Formation and the second member of the Shahejie Formation in the Dongying Sag.

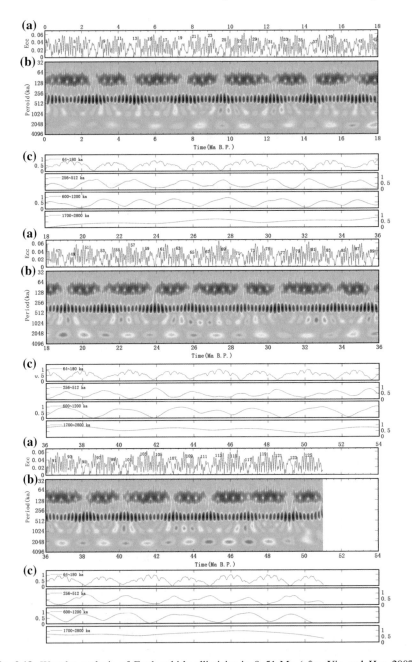

Fig. 3.18 Wavelet analysis of Earth orbit's ellipticity in 0–51 Ma (after Yin and Han 2007). **a** Theoretical time sequence of ellipticity; **b** Amplitude spectrum of wavelet; **c** Normalized average time sequence

3.2.2 Stratigraphic Age and Erosion Amount Calculation Method

1. Calculation rationale.

Stratigraphic age and erosion amount calculation is based on Milankovitch Cyclicity, especially the 400 ka periodic sedimentary cycle change. Stratigraphic thickness of key wells was transformed to time and average geological time can be calculated. Therefore, the Mesozoic–Cenozoic astronomy stratigraphic age chart of certain basin can be established. Then based on multiple wells, stratigraphic age of basin, erosion time and eroded thickness can be calculated, which provides the basis for the quantitative study of hydrocarbon generation and accumulation.

The calculation procedure is: ① to select well logging data (GR, SP, RT) on main cross sections, and relevant seismic, sedimentology, paleontology analysis to confirm thickness, top and bottom of each stratigraphic layer; ② to analyze the frequency spectrum of well logging data and search for the 400 ka long period peak and its main corresponding preferential cycle (m); ③ to compare main preferential cycle with the astronomy theoretical cycle, transform strata thickness to time (ka), and calculate sediment accumulation rate (m/ka) and the gap time of subsidence (Ma); ④ to smooth main preferential cycle of various layers, extract the Milankovitch Cyclicity period and interval time, compare with the astronomy theoretical period, and determine interval age of the layer; ⑤ to take paleomagnetism, paleontology, sediment and seismic reflection characteristics of each well for the calculation of age and geological interval time, determine the absolute age of the well top and bottom; ⑥ to use total strata interval time minus present strata sediment time divided 0.405 Ma and multiply by thickness to get the eroded thickness of the strata.

2. Calculation method.

(i) Frequency spectrum cycle and analysis.

Frequency spectrum analysis is applied via fast Fourier transform (FFT). Frequency spectrum analysis is according to the well logging time sequence to seek hidden cycles and take one time sequence as a group of frequency combination. Any periodic function can use Sine and Cosine wave combination in the approach. During the spectrum analysis, Fourier transform is actually a linear superposition of Sine wave with different frequencies to differentiate wave strength. The Fourier transform function has infinite time and is not localized.

The period unit is time or thickness. Frequency is the reciprocal of the period. The periodic function is determined by FFT, therefore, a power spectrum can be determined. Each frequency component range labeled on a histogram is called a spectrogram, which shows the variable range of frequency. The higher range of frequency has more contribution to the data sequence. In astronomy strata, the

Table 3.2 Comparison among long periodicity values of different nds (measuring unit is m; point distance is 0.125 m)

nd	4096 points	2048 points	1024 points	512 points
1	512			
2	256	256		
3	170.67			
4	128	128	128	
5	102.4			
6	85.33	85.33		
7	73.14			
8	64	64	64	64
9	56.89			
10	51.2	51.2		
11	46.55			
12	42.67	42.67	42.7	
13	39.38			
14	36.57	36.57		
15	34.13			
16	32	32	32	32

Notes Unit for cycles is m; Point distance is 0.125 m

research focuses on low frequency and middle frequency of the frequency spectrum. Frequency spectrum analysis could be completed using ⟨cyclostratigraphy research system⟩ Version1.0.

Commonly used frequency spectrums are GR, SP, and Rt and sample depth interval is 8 points/m (0.125 m).

Point selection is exponential (nd), which are 512, 1024, 2048, and 4096. It is determined by the nature of Fourier analysis. If point distance is 0.125 m, the nd converts into thicknesses of 64, 128, 256, and 512 m. In most cases, spectrum analysis could use dozens of or a hundred meters of well logging data to display the relevant thickness to astronomy 405 ka eccentricity period. Data length and long period resolution by Fourier transform demand nd to be power of 2 (Table 3.2). The long period resolution of 1024 and 512 points is obviously far lower than 4096 and 2048 points.

(ii) Major preferential cycle and accumulation rate calculation.

Frequency spectrum analysis commonly show multi-preferential cycles, but what needs to be calculated is main cycle peak approaching 400 ka to compare with 405 ka astronomical periodicity. Accumulation rate describes the rate of different lithological assemble (including sandstone, shale, and congramerate) build up (m/ka). Accumulation rate (AR) = main preferential cycle discovered in 400 ka (m/405 ka).

(iii) Applied digital filtering to achieve period change.

Frequency or period element in the sedimentary section is provided by using the frequency spectrum, i.e., by using digital filtering to achieve period change, for example, using $x(t)$ to represent observed data, which is the input signal and use $h(t)$ represents filter. Time domain filter is input signal, making certain frequency signal pass while other frequency components were filtered. After certain calculation of the input signals, only the needed signal could be received. The mathematical conduction is called digital filtering if the transform function is performed via algorithm. The frequency range that the signal could go through the filtering is called the passband, while the restrained frequency is called the retardant. Ideal filtering should filter useless signals and output useful signals without distortion; however, it was impossible in reality. The actual property of filtering is only an ideal means to transform well logging data into a period change.

(iv) Using wavelet analysis to transform well logging sequences into time sequence.

Wavelet analysis or wavelet transform is kind of mathematical transform which is extensively applied in signal processing, using limited length or rapid attenuation and wavelet vibration form to represent a signal. Wavelet transform provides one tunable two-dimensional time–frequency window. Morlet's wavelet frequency characteristics are remarkable, and many people use them in cyclostratigraphy research.

The frequency modulation of wavelet analysis has a "target" of the selected 400 ka period main preferential cycle. The time control point and AR of a stratal unit can be determined and well logging data distance (space) can be transformed into time. The wavelet analysis can be performed on evenly distributed time sequence with interpolation. From the wavelet chart, each stratigraphic unit period and AR of the main preferential cycle are derived from the period peak, valley, and relevant 400 ka period position of well logging data sequence.

(v) Continual time and absolute age calculation.

There are two ways to calculate astronomy strata continual time and absolute age. ① The first is to analyze the spectrum of well logging data, seek the 405 ka period cycle peak thickness divided by 405 ka to get AR, and use the top, bottom and thickness of the stratigraphic section divided by AR to get continual time; ② The second is to use filter analysis data, which means using a filtering band-pass value of the strata multiplied by the corresponding theoretical period to get continual time. Using well Niu38 in the Dongying sag as an illustration, the thickness of the middle Es3 member is 213 m. A band-pass filter of 93 and a Milankovitch precession cycle of 22.4 ka yields a preferential cycle of 2.26 m. The continual time is 22.4 ka × 93 = 2.08 Ma.

Once the continual time of the strata is obtained and if important geological age of magnetism or biostratigraphy is available, the absolute age can be easily estimated by plus or minus the continual time.

3.2.3 Eroded Thickness Calculation in the Dongying Sag and Its Application

In order to calculate eroded thickness in the Dongying sag, 10 key wells were initially selected to establish the stratigraphic timescale of the sag based on labeled layers and Milankovitch cycles. Based on this timescale frame, eroded thickness was calculated for about one hundred wells, which provides to the primary control on generation and accumulation research.

1. Establishment of Cenozoic chronological table of the Dongying sag.

(i) Key well selection and analysis from the Dongying sag.

Demarcated wells are selected mainly based on the representativeness of strata and integrity of GR and SP logging data. For example, well Haoke 1 has good representativeness for the Kongdian Formation-lower section of Es3 member. Well Niu38 has complete Es3 and Es1 members. Dongying formation is best represented in well Li1, Guantao-Minghuazhen formations are best represented in well H8 and Pingyuan Formation is represented in well D2-4. The geographic location of 9 key wells is illustrated in Fig. 3.19 and the depth of formation bottom from well logging is listed in Table 3.3.

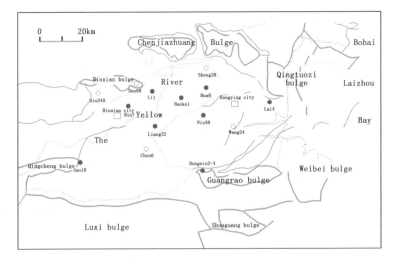

Fig. 3.19 Geographic location of Niu38 and 8 other key wells in the Dongying Sag (modified from Shengli Oilfield 2007)

Table 3.3 Bottom depth (m) of stratigraphic unit in Niu38 and 8 other key wells in the Dongying Sag

Stratum	Gao19	Bin7	Li1	Liang22	Haoke1	Hua8	Dongxin2–4	Niu38	Lai4
4th member of Pingyuan Fm							28 (28)		
3rd member of Pingyuan Fm							161 (133)		
2nd member of Pingyuan Fm							242.5 (81.5)		
1st member of Pingyuan Fm	295	425	252.5	299	300	389	274.5 (32)	396	304
Upper Minghuazhen Fm	552 (257)	745 (315)	643 (390.5)	620 (321)	700 (400)	695 (306)	530.1 (255.6) ▽	695 (299)	
Lower Minghuazhen Fm	836 (304)	1175 (435)	1055 (412)	1156 (536.5)	1065 (365)	1125 (430)		1012 (317)	871 (587)
Upper Guantao Fm	1130.5 (320.5)	1462 (287)	1396 (341)	1650 (493.5)	1328 (263)	1430 (76)			
Lower Guantao Fm					1462 (134)			1350 (338)	1174 (303)
1st member of Dongying Fm	1195 (64.5)	1523 (61)	1404 (8)	1706 (59)	1565 (103)	1479 (49)		1600 (105)	1352 (178)
2nd member of Dongying Fm	1372.5 (177.5)	1763.5 (242.5)	1647.5 (243.5)	2004 (295)	1766 (201)	1714 (235)		1787 (187)	
3rd member of Dongying Fm		2017 (251.5)	1958 (310)	2327 (323)	2027 (261)	1750 (36) ▽		1958 (171)	
Es1	1438.5 (66)	2288.5 (271.5)	2202.5	2551 (224.5)	2173 (146)			2192 (234)	1459.5 (107.5)
Upper Es2		2365 (76.5)	2415	2918 (366.5)	Fault			2300 (108)	1817 (357.5)
Lower Es2	1658 (219.5)	2477 (112)	2513.25					2425 (125)	2072 (255) ▽

(continued)

Table 3.3 (continued)

Stratum	Gao19	Bin7	Li1	Liang22	Haoke1	Hua8	Dongxin2–4	Niu38	Lai4
Upper Es3	1811 (153)	2625.5 (148.5)	3205 ▽	3294 (376.1) ▽	2573 (400)			2724 (299)	
Upper part of Middle Es3		2858 (232.5)			2929 (356)			2952 (228)	
Middle part of Middle Es3								3050 (98)	
Lower part of Middle Es3								3263 (213)	
Upper part of Lower Es3		3211 (353)			3177 (248)			3367 (104)	
Lower part of Lower Es3								3450 (83) ▽	
Upper Es4	2009 (198) ▽	3528 (317) ▽			3398 (221)				
Middle Es4					3755 (357)				
Lower Es4					4714.5 (959.5) ▽				
Upper part of 1st member of Kongdian Fm					5492 (777.5)				
Lower part of 1st member of Kongdian Fm					5807.8 (315.8) ▽				
Logging series	SP, GR	SP, GR	SP, GR	SP	SP, GR	SP	SP, Mi	SP, GR	SP
Magnetic strata							0–530	2770–3267	
Paleontology	1130–1750	1523–2017	1396–1958	2017–3528	2573–5807	389–1750	0–530	2770–3367	1174–2072

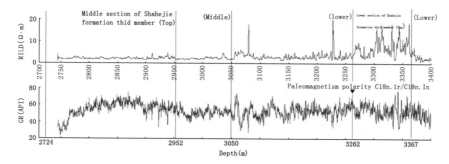

Fig. 3.20 Stratification of middle and lower parts of the Es3 member in well Niu38 based on GR and RILD borehole logs

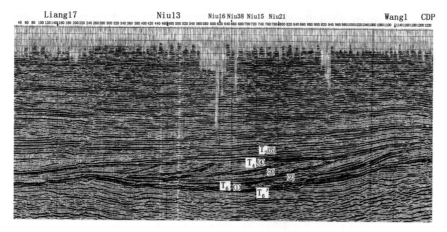

Fig. 3.21 An illustration of seismic reflection characteristics of Line 94.0 in the Dongying sag. ① Lower sub-member of Es3 (T$_6$); ② Lower part of the middle sub-member of Es3; ③ Middle part of the middle sub-member of Es3; ④ Upper part of the middle sub-member of Es3 (T$_4$); ⑤ Es2 + Es1 (T$_2$)

(ii) Stratigraphic age of the middle part of ES3 in well Niu38 calculated by Milankovitch cycles.

(1) Stratigraphic layering.

Based on SP, GR and seismic reflection, fossils, sediment and magnetism strata data of well Niu38, four demarcated layers can be recognized from the middle part of the Es3 member (2774–3367 m) (Figs. 3.20 and 3.21).

(2) Well logging frequency spectrum analysis to find the 405 ka long period peak and main preferential cycle.

The top and bottom depths of the middle and lower part of the Es3 member from GR data at well Niu38 were used as an example. The depth range is from 2901 to 3412.875 m. Using three different lengths of 512 m (4096 points), 256 m (2048

Niu38 well GR Begin depth: 2774m Data point distance=0.125m Length: 2048 point

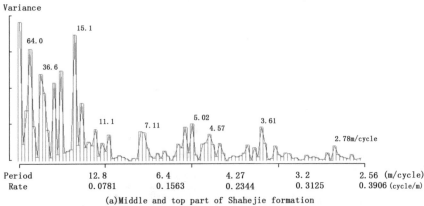

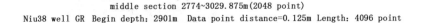

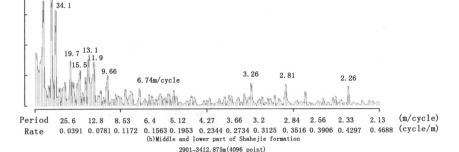

Fig. 3.22 Spectrum of the Shahejie Formation in well Niu38

points) and 128 m (1024 points) to do spectrum analysis, the main preferential cycle can be recognized. Here only spectrograms of the middle and upper part of Es3 member (2774–3029.875 m) with 2048 points and middle and lower part of Es3 member (2901–3412.875 m) with 4096 points are displayed to illustrate the maxima of the main preferential cycle. The 64.0 and 42.7 m maxima correspond to the 405 ka Milankovitch cycle (Fig. 3.22). In addition, the other main preferential cycles are found and calculated (Table 3.4).

Other preferential period cycle thicknesses from Table 3.4 allow the calculations for 405 ka and other periods. For example, the power spectrum of the middle-lower part of Es3 member, provides a main preferential cycle of 42.7 m in response to the long period eccentricity of 405 ka, and an average deposition rate of the cycle of 42.7 m/405 ka = 0.1054 m/ka. A similar method can be used to calculate the main preferential cycle, period and AR. The calculated results can be compared with the theoretical period derived from Milankovitch cycles.

Table 3.4 Main preferential cycles in GR spectrum of well Niu38

Stratum	Depth (top to bottom) (m)	Preferential cycle (m/cycle)
Upper Es3 member	2430–2685.875 (2048 points)	42.7 23.2 2.64 5.69 8.83 2.88 3.77 2.17
Middle Es3 member	2751–3262.875 (4096 points)	34.1 42.7 64.0 28.4 19.7 15.5 12.5 7.11 9.31 5.39
Upper and middle part of middle Es3 member	2774–3029.875 (2028 points)	15.1 64.0 19.7 36.6 5.02 3.61 11.1 7.11 4.57 2.78
Middle and lower part of Es3 member	2901–3412.875 (4096 points)	42.7 85.3 34.1 13.1 19.7 11.9 15.5 9.66 3.26 2.81 2.26 6.74
Middle part of middle Es3 member	2952–3079.875 (1024 points)	16.0 32.0 10.7 6.74 1.73 2.25 3.67 1.42
Middle and lower part of middle Es3 member	3000–3255.875 (2048 points)	12.2 36.6 16.0 8.26 6.74 1.73 2.25 3.67 1.42
Lower part of middle Es3 member (Upper)	3050–3177.875 (1024 points)	42.7 16.0 11.6 8.00 6.10 3.12 4.00 2.25 2.72 1.86
Lower part of middle Es3—Lower Es3 member	3135–3390.875 (2048 points)	42.7 19.7 3.61 9.85 2.81 8.20 6.00 5.12 3.37 2.61 4.41 1.88
Lower part of middle Es3 member (Lower)	3135–3262.875 (1024 points)	42.7 12.8 8.53 2.61 3.37 5.12 1.88 1.42
Upper part of lower Es3 member	3263–3390.875 (1024 points)	42.7 14.2 3.28 5.57 7.11 2.84 2.25 1.83

Table 3.5 Main preferential cycles, periodicity and Milankovitch astronomical cycles in well Niu38

Milankovitch cycle period (ka) (after Hinnov 2005)	Upper and middle part of middle Es3 member		Middle and lower Es3 member	
	Cycle thickness (m)	Period (ka)	Cycle thickness (m)	Period (ka)
			85.5	809.1
405 (eccentricity 1)	64.0	405	42.7	405
			34.1	323.4
	36.6	231.6	19.7	186.9
131 (eccentricity 5)			15.5	147.0
124 (eccentricity 3)	19.7	124.7	13.1	124.3
99 (eccentricity 4)	15.1	95.6	11.9	112.9
95 (eccentricity 2)			9.66	91.6
	11.1	70.2	6.74	63.9
53.6 (obliquity 3)				
41.0 (obliquity 1)	7.11	45.0		
39.6 (obliquity 2)				
	5.02	31.8	3.26	30.9
23.67 (precession 1)	4.57	28.9	2.81	26.7
22.37 (precession 2)	3.61	22.8	2.26	21.4
19.10 (precession 3)	2.78	17.6		
AR (m/ka)	0.1580		0.1054	

The first column in Table 3.5 shows 11 main theoretical values of Milankovitch cycle periods. The long period eccentricity of 405 ka is the most stable orbital parameter, so it can become the scale template of a stratigraphic period. The second and fourth columns are the GR data spectrum calculation listed in Table 3.4 using middle-lower Es3 member as an illustration. The preferential cycle of 2.26 m divided AR of 0.1054 m/ka derives precession of 21.4 ka. Stratigraphic cycles in the second and fourth columns can be transformed into periods listed in the third and fifth columns, which should be compared with the 11 theoretical values listed in column 1. Therefore, the age of each stratigraphic unit in the well can be calculated by using theoretical astronomical values.

(3) Age calculation at a stratigraphic boundary.

Age calculation can be illustrated by using the lower middle Es3 member in well Niu38. Firstly, age demarcated magnetically at the lower middle Es3 member boundary C18n.lr (depth 3263 m) is 38.975 Ma (Fig. 3.20). The stratigraphic thickness from 3263 m upward to 3050 m (top of the lower part of middle Es3 member) is 213 m and AR is 0.1054 m/ka. The interval time (213 m/0.1051 m/ka) is 2.027 Ma. Therefore, the age at 3050 m in the middle Es3 member is 38.975 Ma − 2.020 Ma = 36.955 Ma. The age at the other subunit in the Es3 member can be determined by the same method (Table 3.6).

Table 3.6 Stratigraphic age of the Es3 member subunits in well Niu38

Stratum	Bottom depth (m)	Strata thickness (m)	Interval time (Ma)	Bottom age (Ma)
Upper Es3 member	2724			34.892
Upper part of middle Es3 member	2952	228	1.443	36.375
Middle part of middle Es3 member	3050	98	0.620	36.955
Lower part of middle Es3 member	3263	213	2.020	38.975
Lower Es3 member	3367 (not penetrated)	104	0.99	

(4)   Cenozoic astronomic stratigraphy timescale in the Dongying Sag.

Based on the age calculation of the Es3 member in well Niu38, stratigraphic age can be estimated in wells Gao19, Bin17, Li1, Liang22, Haoke1, Hua8, Dongxin2–4 and Lai4 in the Dongying Sag and the Cenozoic astronomical stratigraphy timescale in the Dongying Sag can be obtained (Table 3.7).

The stratigraphic age of 54.6 Ma at the bottom of the first member of the Kongdian Formation in Table 3.7 contains data of Yao and Xu (2007). Table 3.7 has 4 columns. The first and second columns are adapted from the International Stratigraphic Scale (Gradstein 2004) and Chinese regional stratigraphic scale (National Stratigraphic Council 2002). The third column is adapted from Li (2005) and Yin (2007). Their stratigraphic ages are derived from the Cenozoic astronomic eccentricity of 405 ka period number of 162 (E1–E162) with total continual time of 65.5 ± 0.3 Ma proposed by Laskar et al. (2004). The fourth column shows the stratigraphic ages of the Dongying Sag calculated from Milankovitch cycles. Two paleomagnetism polarity ages of 38.975 Ma at bottom of middle Es3 in well Niu38 (3263 m) and 2.588 Ma at the top of the Minghuazhen Formation in well Dongxin 2–4 (274 m), three main unconformities and eroded thickness are included.

2. Erosion time and eroded thickness analysis of three major unconformities in Cenozoic Dongying Sag.

(i)   The unconformity between lower and upper Es2 member.

There are several uplift and subsidence events in the Shahejie Formation of the Dongying Sag. For example, the boundary between the upper and lower members of the Es2 at 2300 m in well Niu38 shows a sharp difference in lithology. The lower part consists of grey black swamp relict sediment deposited in a shallow lake, while upper part consists of red coarse clastic sediment in a fluvial facies. There is an unconformity between the upper and middle Es2 member, especially in the Shengtuo area. This unconformity becomes an important trap for hydrocarbon accumulations underneath because of the overlapping strata being tight mudstones or shale.

According to the above mentioned astronomical stratigraphy timescale analysis, the age at the bottom of the upper Es3 member is 34.900 Ma, at the bottom of Es2 member is 33.799 Ma, at the bottom of upper Es2 member is 33.338 Ma and at the top of Es2 member is 32.940 Ma. The deposition time of the lower Es2 member is 33.799 Ma − 33.338 Ma = 0.461 Ma and that of the upper Es2 member is 33.338 Ma − 32.940 Ma = 0.398 Ma. The total deposition time is 0.461 + 0.398 = 0.859 Ma, less than 1 Ma. The denudation time should be less than 0.5 Ma. The eroded thickness calculated from 405 ka period 110 m/cycle in well Niu38 can be estimated as 0.5 Ma/0.405 Ma × 110 m = 135 m. The actual erosion time in well Niu38 is 0.038 Ma and the eroded thickness is about 10 m (Table 3.8).

According to the wavelet analysis of well Niu38, the depth at lower/upper boundary of Es2 member is 2425 m and the time is 33.4 Ma. The dominant period of earth orbit shifts from 41 to 100 ka (Fig. 3.23). This means that the eccentricity 405 ka amplitude from the upper Es3 member to lower Es2 member is very high, while 100 ka period amplitude is low and scatter. However, eccentricity 405 ka from the upper Es2 member to the Es1 member returns to a lower value, while the 100 ka period amplitude turns higher and more orderly. The time from 33.8 to 33.4 Ma is coincident with the shifting of 41–100 ka in astronomical orbit period, leading to depositional facies change and unconformity occurrence.

Recently, according to deep ocean geology research, many scientists observed a large temperature drop at 33.6 Ma. For example, Liu (2004) suggested that during the early Oligocene (about 33.5–33.1 Ma), ocean surface and deep water temperatures decrease sharply. The $\delta^{18}O$ value has increased by 1.4 ‰ quickly within 400 ka during the earliest Oligocene (33.6 Ma) and the bottom water temperature has decreased 3–4 °C. The ice sheet began to emerge at eastern Antarctica, the event is called the "early Oligocene ice age incident" (EOGM or Oi-1). The EOGM at about 33.5–33.1 Ma was coincident with changes in lithologies and carbonate contents as well. According to carbonate content change derived from the color index from South Atlantic ODP208 voyage number 1262 station, the eccentricity of 100 and 400 ka periods can be clearly observed on the 30–33.8 Ma wavelet chart, while the 100 and 400 ka periods are vague before 34 Ma (Fig. 3.24). Identically, Coxall et al. (2005) also noticed large variations in carbonate contents and oxygen isotopic values during 33.6–33.7 Ma in the tropical Pacific Ocean (ODP Leg 199, Site 1218) (Fig. 3.25). These deep ocean drilling results are accordant with the occurrence of the unconformity between the upper and lower Es2 member in the Dongying Sag.

(ii) Analysis of the unconformity between Dongying and Guantao formations.

Middle Dongying Formation sediments are underneath the Guantao-Dongying unconformity in the Dongying Sag. The third and second members of the Dongying Formation consist mainly of deltaic light gray sandstones and mudstones. The maximum sediment thickness (up to 700 m) occurs in the middle-western region around wells Liang 22 and Li1. The lake size decreased during the late period of Dongying Formation deposition with only a limited area around wells Haoke1 and

Table 3.7 Cenozoic geological timescale in the Dongying Sag

International stratigraphic table Gradstein et al, 2012			China regional chronostratigraphic 2002			Li et al , 2005	Stratum classification and bottom age of Dongying sag						
Series	Stage	Bottom age (Ma)	Series (Epoch)	Stage	Bottom age (Ma)	405ka Period	Strata			Strata Code	Bottom age (average) (Ma)	Uplift/Erode Denudation condition	Well
							Formation	Member	Section				
Holocene		0.0117	Holocene		0.01		Pingyuan formation	4th		Qp	① 0.0115		
Pleistocene	Upper	0.126	Pleistocene	Sailawusu stage		1~5		3rd			① 0.126		① Dongxin 2.4
	Middle	0.781						2nd			① 0.781		
	Lower	1.806		Zhoukoudian stage				1st			① 0.967		
				Xihemn stage							1.806	Uplift and eroded thickness 130m	
	Gelasian	2.588			2.60	5~10	Minghuazhen formation	Upper		N₂m₂	①2.572(2.588)		
Pliocene	Piacenzian	3.600	Pliocene	Mingen stage							① 3.598		
	Zanclean	5.333		Gaozhuang stage	5.30	10~14		Lower		N₂m₁	②5.003(5.121)		
Miocene	Messinian	7.246	Miocene	Baode stage		14~19				N₁m₁	②11.120 (10.794)		② Hua8
	Tortonian	11.62				19~29							④ Lil
	Serravalian	13.82		Tongga'er stage		30~34	Guantao formation	Upper		N₁g₂	②16.176 (15.393)		⑤ Lai4
	Langhian	15.97		Shanwang stage		35~40		Lower		N₁g₁	②18.095 (17.405)		⑥ Guo19 ⑦ Liang22
	Burdigalian	20.44				41~51							
	Auitanian	23.03	Oligocene	Xiejia stage	23.3	52~58					23.03	Lift and eroded thickness about 10-600m	③ Haoke1
Oligocene	Chattian	28.1		Taben buluke stage		59~64	Dongying formation	1st		E₃d₁	④24.414(24.467) ④24.541 (25.385)		③ Niu38
						64~80		2nd		E₃d₂	④28.393 (28.406)		⑧ Bin7
				Wulanbulage stage	32			3rd		E₃d₃	③31.829		
	Rupelian	33.9		Caijiachong stage		80~83	Shahejie formation	1st	Upper M/L	E₂s₁	③ 32.940	Lift Unconformity	
								2nd	Upper	E₂s₂	③ 33.338		
									Lower		③ 33.799		
	Priabonian	38.0	Eocene	Yuanqu stage		84~93		3rd	Upper U M L	E₂s₃	③ 34.900 ③ 36.275 ③ 36.895 ③ 38.975		
Eocene	Bartonian	41.3			93~101				Lower		⑤ 40.904		
	Lutetian	47.8		Lushi stage	56.5	101~121		4th	Upper Middle Lower	E₂s₄	⑤ 42.671 ⑤ 44.264 ⑤ 48.148		
	Ypresian	56.0		Lingcha stage			Kongdian formation	1st	Upper Lower	E₂k₁	⑤ 50.536 54.638		
								2nd	Upper U Middle L	E₂k₂			
Paleocene	Thanetian	59.2	Paleocene	Chijiang stage		121~162				E₁k₂			
	Selandian	61.6						3rd	Upper Lower	E₁k₁			
	Danian	66.0		Shanghu stage	65.0						66.0		

Table 3.8 Denudation time and eroded thickness between lower and upper Es2 members in well Niu38

Stratum		Depth (m)	Thickness (m)	Cycle (m)	400 ka (Ma)	Interval time (Ma)	Bottom Age (Ma)	Erosion time (Ma)	Eroded thickness (m)
Dongying Fm	Ed	1429–1958	529				31.829		
Es1 member	Es_1	1958–2192	234	118.2	0.405	0.802	32.631		
Upper Es2 member	Es_2^s	2192–2300	108	110.7	0.405	0.395	33.026		
Lower Es2 member	Es_2^x	2300–2425	125	110.7	0.405	0.457		0.038	10
							33.064 33.521		
Upper Es3 member	Es_3^s	2425–2724	299	100.1	0.405	1.210	34.731		
Upper part of middle Es3 member	$Es_3^z(s)$	2724–2952	228	68.45	0.405	1.349	36.080		
Middle part of middle Es3 member	$Es_3^z(z)$	2952–3050	98	50.67	0.405	0.783	36.863		
Lower part of middle Es3 member	$Es_3^z(x)$	3050–3263	213	42.88	0.405	2.012	38.875		
Lower Es3 member	Es_3^x	3263–3450	187	42.84	0.405	1.768	40.643		

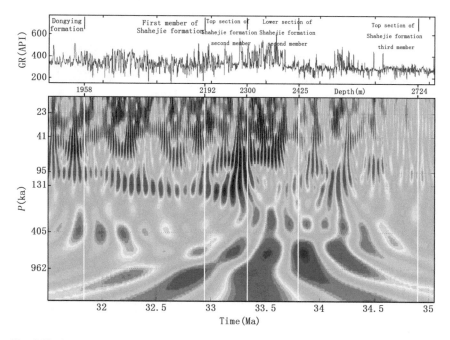

Fig. 3.23 GR logging curve (*upper*) and wavelet graph in the duration of 31.5–35 Ma of Es1, Es2 and upper Es3 members in well Niu38

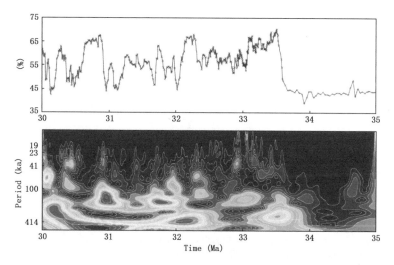

Fig. 3.24 Color reflectance wavelet spectrum in 35–30 Ma of the 208th ODP Leg 208 (Zhifei Li 2004)

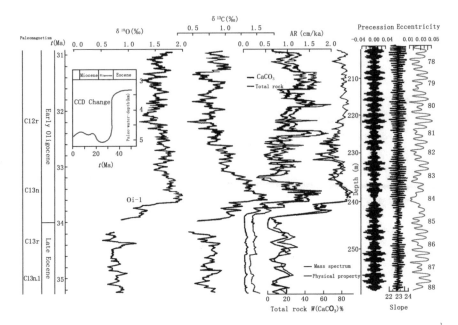

Fig. 3.25 Carbonate accumulation rates and variations of $\delta^{18}O$ and $\delta^{13}C$ values in demersal foraminifers from Eocene to Oligocene at Site 1218 in the tropical Pacific Ocean (adapted from Coxall et al. 2005)

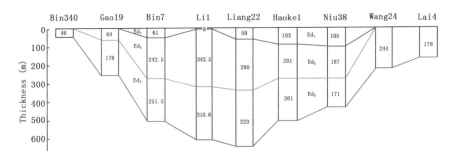

Fig. 3.26 Cross section from well Bin340 to well Lai4 showing current thickness of the Dongying Formation in Dongying Sag

Niu38. The first member of the Dongying Formation consists mainly of fluvial and alluvial facies of brownish red/variegated sandstones and mudstones with maximum thickness of about 100 m (Fig. 3.26). Above the unconformity the Neocene Guantao and Minghuazhen formations entirely covered the sag. Strata under the unconformity vary from the Dongying Formation, Shahejie Formation to pre-Cenozoic. The major faults are not active during the deposition of the Neocene strata. For example, the unconformity between the Guantao Formation and underlying Paleocene or older strata is very clear especially in the Shanjiasi area in

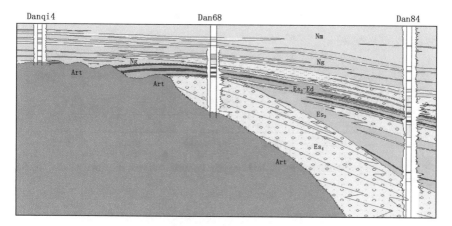

Fig. 3.27 Thickness variation of Minghuazhen and Guantao formations in the Dongying Sag

western Dongying sag (Fig. 3.27). The thickness of Guantao Formation in Dongying area is about 200–500 m, while Minghuazhen Formation is flat with more silty mudstone and has thickness of about 500–1000 m.

In order to calculate the denudation time and eroded thickness of the Dongying/Guantao unconformity, three main geological ages were investigated. ① As the international Oligocene/Miocene age is calculated according to E58 long period eccentricity's lowest amplitude of 100 ka period earth orbit (Lourens et al. 2004; Gradstein et al. 2005), Dongying/Guantao unconformity time in the Dongying Sag is 23.03 Ma. ② According to the astronomical stratigraphy time table, the age at bottom of the Dongying Formation is 31.829 Ma (Table 3.7), but according to Chinese regional chronostratigraphy (2002), the time at the bottom of the Dongying Formation in Bohai Bay Basin is 32 Ma, which is based on the weakest amplitude of E80 in 405 ka period (Yin and Han 2007). We picked 32 Ma as age of bottom Dongying Formation. ③ Due to variable top and bottom Guantao Formation in the Dongying Sag, we used the top of the Minghuazhen Formation with known age of 2.588 Ma to calculate downward for time of the Guantao Formation.

The eroded thickness of the Dongying Formation has been calculated for nine wells. Bin7 has been used as an illustration (Fig. 3.19). Dongying Formation from 1462 to 2017 m has the thickness of 555 m. Spectrum analysis shows the main preferential cycle of 32.79 m and 405 ka period and time for deposition can be determined as 555 m/32.79 m × 0.405 Ma = 6.8560 Ma. As the age at bottom of the Dongying Formation is 32 Ma, the age at top boundary before erosion is 32 Ma − 6.8560 Ma = 25.145 Ma. Therefore, the denudation time of Dongying/Guantao unconformity is 25.145 Ma − 23.03 Ma = 2.115 Ma and the eroded thickness is 2.115 Ma/0.405 Ma × 32.79 m = 171 m (Figs. 3.28 and 3.29).

Using the same method as that for well Bin7, the denudation time and eroded thickness of the Dongying/Guantao unconformity for the other eight wells in the Dongying Sag were calculated and are listed in Table 3.9.

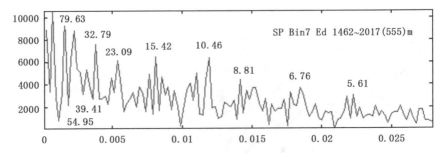

Fig. 3.28 Major preferential cycles of the Dongying Formation at well Bin7 in the Dongying Sag (m/cycle)

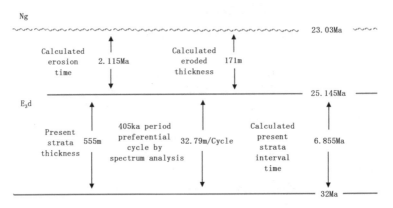

Fig. 3.29 Illustration of denudation time and eroded thickness calculation of the Dongying Formation at well Bin7 in Dongying Sag

The method of the denudation time and eroded thickness calculation are expanded here. ① The maximum denudation time of the Dongying Formation is 32 Ma − 23.03 Ma = 8.97 Ma, where no Dongying Formation strata have been preserved. The erosion time at well Bin340 in western Dongying Sag is 8.50 Ma and the eroded thickness is 839 m. The erosion time at well Lai4 in eastern Dongying Sag is 6.96 Ma and the eroded thickness is 615 m. The erosion time at central and western area is minor. For instance the erosion time at well Liang22 is 0.05 Ma and the eroded thickness is only 3 m. The erosion time at well Haoke1 is 0.67 Ma and the eroded thickness is 42 m. The erosion time at well Li1 is 1.57 Ma and the eroded thickness is 98 m. The erosion time at well Niu38 is 1.36 Ma and the eroded thickness is 99 m. ② Although the deposition of the Guantao Formation above the unconformity starts from the Miocene at 23.03 Ma, the hills remain as heritage highs until about 18–15 Ma when the flat shaped basin at wells Liang22, Haoke1 and Niu38 s area start subsidence. The middle and upper Guantao Formation is widely distributed. The strata dominated by fluvial facies coarse clastic sediments form very good reservoirs. Deposition of the Minghuazhen

Table 3.9 Stratigraphic time of Dongying Formation-Minghuazhen Formation of 9 wells in the Dongying Sag

Strata	Bin340 Top Bottom (m)	Thickness (m)	Cycle thickness (m)	Interval time (Ma)	Strata age (Ma)	Gao19 Top Bottom (m)	Thickness (m)	Cycle thickness (m)	Interval time (Ma)	Strata age (Ma)	Bin7 Top Bottom (m)	Thickness (m)	Cycle thickness (m)	Interval time (Ma)	Strata age (Ma)
					1.806					1.806					1.806
					(1.398)					(2.438)					(1.618)
Nm	185.0				3.204	295.0				4.244	425.0				3.424
		583.5	29.78	7.916	11.120		560.5	32.93	6.876	11.120		750.0	39.37	7.696	11.120
	768.5					855.5					1175.0				
Ng	1050.0	281.5	23.36	4.868	15.988	1130.5	275.0	26.35	4.216	15.336	1462.0	287.0	28.90	4.012	15.132
					(7.042)					(7.694)					(7.898)
					23.03					23.03					23.03
Ed₁					(8.499)					(4.796)					(2.473)
											1523.0	61.0	21.50	1.149	25.503
Ed₂						1195.0	64.5	21.56	1.212	27.826	1765.0	242.5	40.22	2.442	26.295
Ed₃		46.5	40.00	0.471	31.529		177.5	24.27	2.962	29.038		251.5	31.21	3.264	28.736
	1096.5				32	1372.5				32	2017.0				32

Strata	Li1 Top Bottom (m)	Thickness (m)	Cycle thickness (m)	Interval time (Ma)	Strata age (Ma)	Liang22 Top Bottom (m)	Thickness (m)	Cycle thickness (m)	Interval time (Ma)	Strata age (Ma)	Haoke1 Top Bottom (m)	Thickness (m)	Cycle thickness (m)	Interval time (Ma)	Strata age (Ma)
					1.806					1.806					1.806
					(0.054)					(0.982)					(0.835)
Nm	252.5				1.860	299.0				2.788	300.0				2.641
		802.5	35.01	9.260	11.120		857.5	41.58	8.332	11.120		765.0	36.45	8.479	11.120
	1055.0					1156.5					1065.0				
Ng	1396.0	341.0	28.48	4.837	15.957	1650.0	493.5	28.90	6.899	18.019	1462.0	397.0	26.54	6.043	17.163
					(7.073)					(5.011)					(5.867)
					23.03					23.03					23.03
Ed₁	1404.0				(1.555)					(0.045)					(0.668)
		8.0	25.60	0.127	24.585	1709.0	59.0	29.99	0.797	23.075	1565.0	103.0	25.60	1.629	23.698
Ed₂	1647.5	243.5	25.60	3.852	24.712	2004.0	295.0	29.34	4.072	23.872	1766.0	201.0	32.00	2.544	25.327
Ed₃		310.5	36.60	3.436	28.564		323.0	32.25	4.056	27.944		261.0	25.60	4.129	27.871
	1958.0				32	2327.0				32	2027.0				32

Strata	Niu38 Top Bottom (m)	Thickness (m)	Cycle thickness (m)	Interval time (Ma)	Strata age (Ma)	Wang24 Top Bottom (m)	Thickness (m)	Cycle thickness (m)	Interval time (Ma)	Strata age (Ma)	Lai4 Top Bottom (m)	Thickness (m)	Cycle thickness (m)	Interval time (Ma)	Strata age (Ma)
					1.806					1.806					1.806
					(1.566)					(1.628)					(1.374)
Nm	396.0				3.362	302.5				3.434	304.0				3.180
		616.0	32.08	7.758	11.120		645.5	33.93	7.686	11.120		587.0	28.85	8.220	11.120
	1012.0					948.0					871.0				
Ng	1429.0	417.0	26.13	6.447	17.567	1347.0	399.0	28.48	5.660	16.780	1174.0	303.0	17.82	6.869	17.989
					(5.463)					(6.250)					(5.041)
					23.03					23.03					23.03
Ed₁					(1.361)					(4.673)					(6.956)
	1600.0	171.0	29.34	2.360	24.3908										
Ed₂	1787.0	187.0	31.21	2.427	26.7512										
Ed₃		171.0	24.54	2.822	29.1779		244.0	23.00	4.297	27.703		178.0	35.8	2.014	29.986
	1958.0				32	1302.5				32	1352.0				32

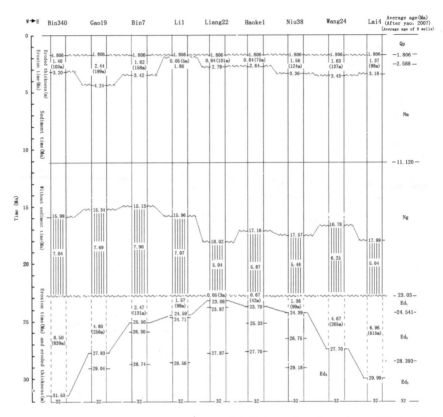

Fig. 3.30 Correlation of geological age and eroded thickness from Dongying to Minghuazhen formations at well Haoke1 and 8 other wells

Formation starts from 10.7 Ma, which is dominated by mudstone and siltstone. The whole basin becomes graben shaped prairie and forms a regional cap rock for oil accumulations. The time boundary of the 405 ka orbital maxima (11.120 Ma and 10.794 Ma) based on palaeoenvironment analysis of regional Guantao and Minghuazhen formations and the frequency spectrum and wavelet at wells Hua8 and Haoke1 is consistent with the Miocene age boundary from the international stratigraphic Table (11.608 Ma) (Table 3.7 and Fig. 3.30).

Pacific Rim drilling data confirmed that water temperature of the ocean bottom reaches a maximum value for the Neogene during the early and middle stage of middle Miocene (16 Ma) and the temperature difference between polar and tropic regions is minimum. This is called "middle Miocene warm period" and lasted for 4 Ma (17.5–13.5 Ma) before severe cooling started (Jin et al. 1995).

It worth mentioning that the occurrence of Shangwang fresh water kieselguhr located about 120 km south of the Dongying Sag and the deposition of late Guantao Formation (13.11 Ma) is the same age as the diatomaceous Monterey Formation in California USA, which is one of the most important source rocks (Fig. 3.31).

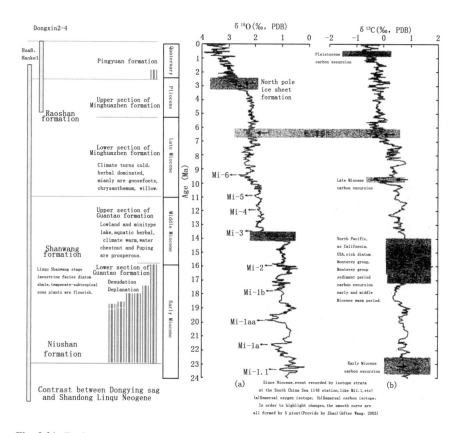

Fig. 3.31 Environmental variation during 16–5.33 Ma and before and after this time period with data integrated from the Dongying Sag, Shandong Linju County and ocean

(ii)   Analysis of Minghuazhen/Pingyuan formation unconformity.

Unconformity analysis of Minghuazhen/Pingyuan formation is mainly on the basis of paleomagnetism of 452 m of core from well Dongxin2–4 in southern Dongying Sag. Four polarity chrons of Brunhes, Matuyama, Gauss and Gilbert were obtained (Fig. 3.32).

Stratigraphic intervals at well Dongxin2–4 from top to bottom are Pingyuan Formation 0–274.5 m, upper Minghuazhen Formation 274.5–452.3 m, lower 452.3–530.1 m (not penetrated). Magnetic stratigraphic ages are: Brunhes polarity chron age of 0.781 Ma at depth 241.5 m and Gauss polarity chron age of 3.596 Ma at depth 452 m. The magnetization intensity curve was used to analyze the frequency spectrum and 404 ka period main preferential cycle was selected for the calculation. ① 0–274.5 m (Pingyuan Formation + top of Minghuazhen Formation) has 102.4 m/cycle and AR is 0.2535 m/ka; ② 274.5–530.1 m (upper Minghuazhen

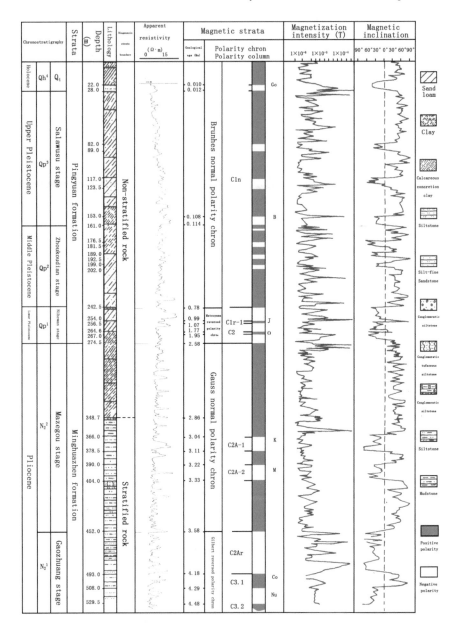

Fig. 3.32 Integrated magnetostratigraphic histograph of well Dongxin2–4

Formation) has 64.0 m/cycle and AR is 0.1584 mm/a. The calculated ages from upper Minghuazhen Formation to the first member of Pingyuan Formation are listed in Table 3.10.

Table 3.10 Geological time from Upper Minghuazhen Formation to the first member of the Pingyuan Formation at well Dongxin2–4

Stratum	Bottom depth (m)	Thickness (m)	Astronomic strata calculation data			Paleomagnetism data
			AR (m/ka)	Continual time (ka)	Age (Ma)	(Gradstein et al. 2005) (Ma)
2nd member of Pingyuan Fm	241.5				0.781	
1st member of Pingyuan Fm	254.0	12.5	0.2535	45	0.826	Kamikatssura (0.886)
	256.5	2.5		10	0.836	
	264.6	8.1		32	0.868	Santa Roso (0.922)
	267.0	2.4		10	0.878	
	274.5	7.5		30	0.908	
	274.5	No sediment		897	1.806	Jaramillo (0.988–1.072)
Upper Minghuazhen Fm		Thickness missing	0.16892	740	2.546	Otway (1.788–1.945) —1.806— Reunion (2.128–2.148) (2.581) Bottom of Matuyama
	348.7	74.2		439	2.985	Diagenesis/not diagenesis boundary of this area
	366.0	17.3		102	3.087	Top of Kaena (3.032)
	378.5	12.5		74	3.161	Bottom of Kaena (3.116)
	390.0	11.5		68	3.229	Top of Mammoth (3.207)
	404.0	14.0		83	3.312	Bottom of Mammoth (3.330)
	452.0	48.0		284	3.596	Bottom of Gauss (3.596)

① Referring to magnetostratigraphical time scale; ② Singer et al. (1999)

Table 3.11 shows age calculation using data of Pingyuan Formation as an example. The time range of the stratigraphic interval of 241.5–254.0 m with thickness of 12.5 m and 102.4 m/cycle can be calculated as: 12.5 m/102.4 m × 0.404 Ma = 0.049 Ma. Brunhes polarity chron age at the top boundary of 241.5 m as determined by magnetic strata is 0.781 Ma. Therefore, the age at the top of the first member of Pingyuan Formation is obtained by adding continual time of 0.049 Ma to be 0.830 Ma.

The top magnetic strata Brunhes polarity chron age of 0.781 Ma at 241.5 m and bottom Gauss polarity chron age of 3.596 Ma at 452 m were used in the calculation. There is an unconformity at 274.5 m. The erosion time from 2.588 Ma to 0.908 Ma is 1.68 Ma.

Table 3.11 Cenozoic astronomical stratigraphy duration and eroded thickness

Stratum				Bottom age (Ma)	Time interval (Ma)	Uplift and eroded thickness
Q	Pingyuan Fm			1.806	2.588	Uplift with about 100 m erosion
N	Minghuazhen Fm	Upper		5.003	2.415	
		Lower		11.120	6.117	
	Guantao Fm	Upper		16–18	11.910	
		Lower		23.03		Uplift with 10–600 m erosion
E₃	Ed1			32	8.970	
	Ed2					
	Ed3					
E₂	Es1			32.940	0.940	
	Es2	Upper		33.338	0.859	
		Lower		33.799		Uplift with about 135 m erosion
	Es3	Upper, Middle		38.750	7.105	
		Lower		40.904		
	Es4	Upper		48.148	7.244	
		Middle				
		Lower				
	Ek1			54.638	6.490	
	Ek2	Upper, Middle		56.689	6.464	
E₁		Houzhen Fm	Middle, Lower	61.102		
	Ek3			66	4.398	

Based on sedimentology and paleontology analysis at well Dongxin2–4, the top of Minghuazhen Formation at 274.5 m is coincident with the transition from continental to marine deposition. The age of Quaternary bottom is 1.806 Ma on the basis of international stratigraphic scale. Meanwhile, ages can also be calculated from astronomical strata. ① Age at the current top of the Minghuazhen Formation is 2.546 Ma and the bottom of Quaternary is 1.806 Ma, therefore, the erosion time is 2.546 Ma − 1.806 Ma = 0.74 Ma. The eroded thickness can be calculated on the base of 404 ka cycle of 102.4 m, which is 0.74 Ma/0.404 Ma × 102.4 m = 188 m; ② The Quaternary bottom of 1.906–0.908 Ma includes time interval of 1.628 Ma which has no deposition due to tectonic movement and uplift. The deposition starts from the first member of the Pingyuan Formation currently buried at depth of 274.5 m with age of 0.908 Ma. This process carries on until the present surface.

More than 100 m of sediment have probably been deposited during last 1.6 Ma on the base of boundary between Minghuazhen and Pingyuan formations (Pliocene and Pleistocene) at well Dongxin2–4. However, these strata are largely eroded due to tectonic movement and uplift, which exerts influences on pore expansion and compaction of reservoirs and finalizes sediment unloading and hydrocarbons accumulation.

3. Eroded thickness calculation of the Cenozoic in the Dongying Sag.

(i) Cenozoic stratigraphic time intervals in the Dongying Sag.

Age at the bottom of each stratigraphic unit, time interval, tectonic uplift and eroded thickness in the Dongying Sag can be obtained from the above mentioned Cenozoic astronomical stratigraphic scale and three important unconformities (Table 3.11). The table is mainly applied to denudation time and eroded thickness calculation for Dongying, Shahejie and Kongdian formations. The time interval of 66–54.638 Ma covers strata deposited in the adjacent areas, which are included in the table to show entire Cenozoic.

(ii) Present sediment in the Dongying Formation and eroded thickness calculation for various strata.

(1) Present sediment and erosion after deposition.

Dongying Sag has received various sediments during the Paleocene starting with the Kongdian Formation, Shahejie Formation to Dongying Formation. The most important unconformity occurs at 23.03 Ma. Before deposition of the Guantao Formation, various strata have been eroded. Strata underneath the Guantao Formation might be Dongying Formation whose bottom age is 32 Ma, Es1 member whose bottom age is 32.94 Ma, Es2 member whose bottom age is 33.799 Ma, Es3 member whose bottom is 40.904 Ma, Es4 member whose bottom age is 48.148 Ma or Kongdian Formation (Fig. 3.33).

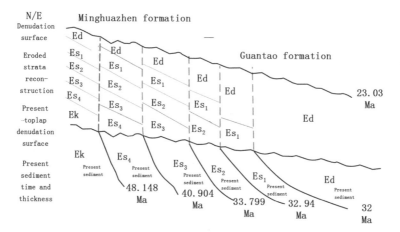

Fig. 3.33 Illustration of denudation time and eroded thickness calculation at Paleocene/Miocene unconformity (23.03 Ma) in the Dongying Sag

Table 3.12 Denudation time and eroded thickness of the Dongying Formation at 142 wells in the Donging Sag

Well	Formation top (m)	Formation bottom (m)	Formation thickness (m)	405 ka cycle (m)	Sediment deposition time (Ma)	Erosion time (Ma)	Eroded thickness (m)
Liang22	1650.00	2327.00	677.00	30.72	8.925	0.045	3
Tong6	1663.00	2313.50	650.50	29.60	8.900	0.070	5
Niu107	1501.00	2051.00	550.00	26.19	8.505	0.465	30
Bin440	1473.00	2153.00	680.00	32.44	8.490	0.480	38
Hao K1	1462.00	2027.00	565.00	27.58	8.297	0.673	46
Wang78	1503.00	2038.50	535.50	26.91	8.059	0.911	61
Liang217	1519.50	2089.00	569.50	28.48	8.099	0.871	61
Bin408	1399.00	2097.50	698.50	35.01	8.080	0.890	77
Liang102	1463.00	2023.00	560.00	29.34	7.730	1.240	90
He132	1377.00	1874.00	497.50	26.91	7.487	1.483	99
Niu38	1429.00	1958.00	529.00	28.50	7.517	1.453	102
Ying93	1373.00	1870.00	497.00	27.30	7.373	1.597	108
Li1	1396.00	1958.00	562.00	30.25	7.524	1.446	108
Bin444	1582.00	2152.00	570.00	30.72	7.515	1.455	110
Bin100	1321.50	1840.00	518.50	28.90	7.266	1.704	122
Hao3	1364.00	1898.00	534.00	29.78	7.262	1.708	126
Bin107	1380.00	1931.50	551.50	30.71	7.273	1.697	129
DongF8	1426.00	2050.00	624.00	34.53	7.319	1.651	141
Dong1	1421.00	1901.50	480.50	28.07	6.933	2.037	141
Tuo52	1443.00	1901.00	458.00	27.40	6.770	2.200	149
Tong52	1400.50	1813.50	413.00	25.88	6.463	2.507	160
Ying35	1393.00	1942.20	549.20	32.50	6.844	2.126	171
Bin7	1462.00	2017.00	555.00	32.78	6.857	2.113	171
Bo3	1338.00	1663.00	325.00	23.25	5.661	3.309	190
Bin169	1371.00	2010.00	639.00	37.88	6.832	2.138	200
Liang208	1527.00	1990.00	463.00	30.00	6.251	2.720	201
Tuo26	1315.00	1617.00	302.00	22.80	5.364	3.606	203
DongF1	1301.50	1720.00	418.50	28.07	6.038	2.932	203
Chun61	1366.00	1799.00	433.00	28.90	6.068	2.902	207
Fan2	1359.00	1902.00	543.00	33.93	6.481	2.489	208
Wang79	1480.00	1950.00	470.00	31.21	6.099	2.871	221
Yong97	1435.50	1950.00	514.50	33.93	6.141	2.829	237
Fan4	1367.00	1878.00	511.00	33.80	6.123	2.847	238
Tuo97	1405.00	1659.00	254.00	23.09	4.455	4.515	257
Tuo147	1476.00	1838.50	362.50	28.07	5.230	3.740	259
Tuo148	1458.00	1829.00	371.00	28.48	5.276	3.694	260
Wang24	1302.50	1547.00	244.50	23.00	4.305	4.665	265
Hua3	1026.50	1354.50	328.00	26.91	4.936	4.034	268
JinQ7	1026.50	1354.50	328.00	26.91	4.936	4.034	268

(continued)

Table 3.12 (continued)

Well	Formation top (m)	Formation bottom (m)	Formation thickness (m)	405 ka cycle (m)	Sediment deposition time (Ma)	Erosion time (Ma)	Eroded thickness (m)
Shi128	1518.00	2067.00	489.00	34.53	5.735	3.235	276
Gao19	1130.50	1372.50	242.00	23.50	4.171	4.799	278
Tuo715	1522.00	1856.00	334.00	27.70	4.883	4.087	280
Xin10	1180.00	1579.00	399.00	30.72	5.260	3.710	281
He122	1450.00	1736.00	286.00	25.84	4.483	4.487	286
Guan116	1541.50	2006.00	464.50	33.93	5.544	3.426	287
Tuo93	1475.50	1816.00	340.50	28.48	4.842	4.128	290
FanS1	1499.50	2000.50	501.00	35.8	5.668	3.302	292
Tong39	1314.50	1666.00	351.50	29.34	4.852	4.118	298
Li2	1314.50	1767.50	453.00	34.00	5.396	3.574	300
Fan3	1441.00	1860.00	419.00	32.55	5.213	3.757	302
Hu10	1359.50	1500.00	140.50	20.01	2.844	6.126	303
Hu5	1171.50	1680.00	508.50	36.70	5.612	3.358	304
Shi111	2009.00	2262.00	253.00	25.17	4.071	4.899	304
Tuo79	1466.00	1879.00	413.00	32.60	5.131	3.839	309
Tuo741	1425.00	1863.00	438.00	33.93	5.228	3.742	313
Fan156	1392.00	1892.00	437.00	33.93	5.216	3.754	314
Fan150	1322.00	1772.00	450.00	34.53	5.278	3.692	315
Tuo719	1492.00	1841.00	349.00	30.00	4.712	4.259	315
Chun43	1503.00	1950.00	447.00	34.53	5.243	3.727	318
Tuo3	1425.00	1795.00	370.00	31.21	4.801	4.169	321
Yong11	1302.50	1422.00	119.50	20.22	2.394	6.576	328
Chun1	1332.50	1661.00	328.50	29.78	4.468	4.502	331
Niu30	1487.00	1814.50	327.50	29.78	4.454	4.516	332
Wang46	1300.00	1591.00	291.00	28.40	4.150	4.820	338
Feng142	1467.00	1707.00	240.00	26.19	3.711	5.259	340
Ying20	1463.00	1885.00	422.00	34.50	4.954	4.016	342
Tuo120	1453.00	1862.00	409.00	33.93	4.882	4.088	342
Tuo29	1434.00	1771.00	337.00	30.70	4.446	4.524	343
Ying10	1448.00	1785.00	337.00	30.72	4.443	4.527	343
Yong10	1360.50	1436.00	75.50	19.03	1.607	7.363	346
Ning1	1427.00	1733.20	306.20	30.00	4.134	4.836	358
Tong4	1456.50	1806.00	349.50	32.25	4.389	4.581	365
Tuo77	1425.00	1849.00	424.00	35.80	4.797	4.173	369
Jin34	967.00	1177.50	210.50	26.19	3.255	5.715	370
Tuo73	1413.00	1712.00	299.00	30.70	3.944	5.026	381
Fan147	1330.00	1658.00	328.00	32.79	4.051	4.919	398
Tuo132	1291.00	1584.00	293.00	31.21	3.802	5.168	398
Tuo142	1424.00	1744.00	320.00	32.50	3.988	4.982	400
Zheng25	1144.00	1233.00	89.00	22.30	1.616	7.354	405
Bin638	1380.50	1712.00	331.50	33.35	4.026	4.944	407

(continued)

Table 3.12 (continued)

Well	Formation top (m)	Formation bottom (m)	Formation thickness (m)	405 ka cycle (m)	Sediment deposition time (Ma)	Erosion time (Ma)	Eroded thickness (m)
Tuo713	1472.00	1775.00	303.00	32.30	3.799	5.171	412
Xin53	1424.00	1796.20	372.20	35.50	4.246	4.724	414
Dan17	1012.00	1103.50	91.20	22.82	1.619	7.351	414
Tuo143	1413.00	1768.00	355.00	35.00	4.108	4.862	420
Tuo5	1373.25	1791.00	417.75	37.88	4.466	4.504	421
Tuo149	1431.50	1829.00	397.50	37.20	4.328	4.462	426
Hua8	1464.00	1755.88	291.88	32.50	3.637	5.333	428
Tuo103	1471.50	1821.50	350.00	35.15	4.033	4.937	429
Tong25	1394.00	1661.00	267.00	31.50	3.433	5.537	431
Chen55	1147.00	1243.00	96.00	23.94	1.624	7.346	434
Li6	1184.00	1361.30	177.30	27.67	2.595	6.375	436
Tuo125	1396.00	1680.00	284.00	32.80	3.507	5.463	442
Tuo2	1496.50	1871.50	375.00	37.16	4.087	4.883	448
Li1	1222.00	1392.00	170.00	28.00	2.459	6.511	450
Tuo53	1441.00	1571.00	130.00	26.20	2.010	6.960	450
Tuo82	1283.50	1382.50	99.00	24.85	1.613	7.357	451
Zheng20	1248.00	1289.50	41.50	22.30	0.754	8.216	452
Wang17	1156.00	1253.00	97.00	24.90	1.578	7.392	454
Cao9	1116.50	1291.00	174.50	28.48	2.481	6.489	456
Yong21	1260.00	1333.00	73.00	24.00	1.232	7.738	459
Niu1	1633.50	2010.50	377.00	37.88	4.031	4.939	462
Tong33	901.00	990.00	89.00	24.90	1.448	7.522	462
Sheng1	1349.00	1522.60	173.60	28.90	2.433	6.537	466
Li3	1190.00	1413.00	223.00	31.21	2.894	6.076	468
Tuo761	1398.00	1782.00	384.00	38.60	4.029	4.941	471
Zheng404	1184.50	1202.60	18.10	22.30	0.329	8.641	476
Tuo170	1480.00	1876.00	396.00	39.41	4.070	4.900	477
Hua14	1029.50	1136.00	107.00	26.54	1.633	7.337	481
Tuo7	1418.00	1637.00	219.00	32.00	2.772	6.198	490
Sheng28	1411.60	1641.00	229.40	32.80	2.833	6.137	497
Lai104	1111.50	1224.00	112.50	27.67	1.647	7.323	500
Bin18	1202.50	1433.50	231.00	33.09	2.827	6.143	502
Guan16	1168.50	1308.50	140.00	29.34	1.933	7.037	510
Yong19	1319.00	1554.50	235.50	33.93	2.811	6.159	516
Hua17	1020.00	1311.00	291.00	36.47	3.232	5.738	517
Liu4	1119.00	1272.00	153.00	30.72	2.017	6.953	527
Yan4	1334.00	1489.50	155.50	31.21	2.018	6.952	536
Ning3	1430.00	1683.00	253.00	35.80	2.862	6.108	540
Wang35	1251.00	1409.00	158.00	31.72	2.017	6.953	545
Ning10	1315.00	1625.00	310.00	38.63	3.250	5.720	546
Lai2	1211.50	1371.50	160.00	32.25	2.009	6.961	554

(continued)

Table 3.12 (continued)

Well	Formation top (m)	Formation bottom (m)	Formation thickness (m)	405 ka cycle (m)	Sediment deposition time (Ma)	Erosion time (Ma)	Eroded thickness (m)
Tong9	977.50	1003.00	25.50	26.19	0.394	8.576	555
Bin132	1325.00	1585.00	260.00	37.16	2.834	6.136	563
Lai5	864.00	1074.50	210.50	35.15	2.425	6.545	568
Bin12	1320.00	1639.00	319.00	40.22	3.212	5.758	572
Lai6	790.50	1007.00	216.50	35.80	2.449	6.521	576
Tuo128	1427.00	1639.00	212.00	35.80	2.398	6.572	581
Li37	1256.00	1425.50	169.50	33.93	2.023	6.947	582
Bin512	1125.00	1160.00	35.00	28.00	0.506	8.464	585
Jin1	918.50	972.00	54.50	28.90	0.764	8.206	586
Bin325	1313.00	1499.00	186.00	35.60	2.116	6.854	602
Lai4	1174.00	1352.00	178.00	35.80	2.014	6.956	615
ShengB1	1431.50	1601.50	170.00	35.80	1.923	7.047	623
FengS1	1354.50	1537.50	183.00	36.47	2.032	6.938	625
Jiao11	1230.00	1250.00	20.00	30.00	0.270	8.700	644
Bin334	1340.00	1515.00	175.00	37.00	1.916	7.054	644
Tuo177	1382.00	1462.00	80.00	32.79	0.988	7.982	646
Bin540	1156.00	1202.00	46.00	32.00	0.582	8.388	663
Bin304	1243.00	1350.00	107.00	35.80	1.210	7.760	686
Lai110	1373.00	1524.00	151.00	37.88	1.614	7.356	688
Bin31	1371.50	1382.0	10.50	33.00	0.129	8.841	720
Bin340	1050.00	1096.50	46.50	40.00	0.471	8.499	839

(2) Eroded thickness calculation at Paleocene/Miocene unconformity.

The calculated eroded thickness includes the Dongying Formation and pre-Dongying Formation. Eroded thickness calculations were made for 142 wells selected for the Dongying Formation and 50 wells for pre-Dongying Formation.

There are six steps to calculate the eroded thickness of the Dongying Formation: ① obtain residual strata thickness (m) by using formation top minus formation bottom depth of each well; ② use the 405 ka period main preferential cycle thickness (m) based on a frequency spectrum analysis of well logs; ③ obtain residual strata deposition time (Ma) by using strata thickness of each well divided by the main preferential cycle, and multiply 0.405 Ma; ④ the original deposition time of 8.98 Ma is calculated by using 32 Ma at the bottom of the Dongying Formation minus 23.03 Ma at top of the Dongying Formation; ⑤ determine erosion time (Ma) by using 8.98 Ma minus the residual stratigraphy deposition time; ⑥ obtain erosion thickness by using erosion time of each well divided by 0.405 and then multiplied by main preferential cycle in 405 ka (Table 3.12).

Data illustrated in Table 3.13 show variable eroded thickness. The minimum erosion time of 0.045 Ma with eroded thickness of 3 m occurs at the middle west of the Dongying Sag near well Liang22; while the maximum erosion time of

Table 3.13 Basic information of sandstone sample from well Yan100 at 1715.13 m

Well	Sample #	Stratum	Depth (m)	Lithology	Current porosity (%)	Current permeability (md)
Yan100	389	Es$_3$	1715.13	Sepia oil immersed fine grained sandstone	19.94	422

8.499 Ma with maximum eroded thickness of 839 m occurs far west near well Bin340. The depression center near well Laing22 can be regarded continuous deposition during 32–23.03 Ma time period, while basin margin such as well Bin340 has long erosion time with eroded thickness as high as 839 m.

3.3   The Method and Application of Sandstone Relaxation Calculation

3.3.1   The Principle of Sandstone Relaxation and Measured Results

Sandstone relaxation caused by overlying strata erosion and pressure reduction reflects sandstone elastic characteristic, which can be calculated based on elasticity of sandstone or simulated in laboratory. However, as strata uplift and erosion is a very slow process and diagenetic evolution of sandstone may be continuous during such slow process, the elastic characteristic of sandstone might be interfered by tectonic stress, leading inelastic deformation. If we make assumption that the influence of continuous diagenesis and tectonic stress during uplift can be ignored, the calculation of relaxation volume is an approximation at best.

Rock elastic compression coefficient can be defined as:

$$\alpha_{弹} = -\frac{1}{V} \cdot \frac{dV_{弹}}{d\sigma}$$

The transformed formula is:

$$\frac{dV_{弹}}{V} = -\alpha_{弹}\, d\sigma$$

To do a definite integral at both sides;

$$\ln V \Big|_{V_1}^{V_2} = -\int_{\sigma_1}^{\sigma_2} \alpha_{弹} d\sigma$$

It can derives:

$$\frac{V_2}{V_2} = \exp\left(-\int_{\sigma_1}^{\sigma_2} \alpha_{弹} \, d\sigma\right)$$

$$\frac{\Delta V}{V_1} = \frac{V_2 - V_1}{V_1} = \exp\left(-\int_{\sigma_1}^{\sigma_2} \alpha_{弹} \, d\sigma\right) - 1$$

When valid pressure changed $\Delta\sigma$, the variation of rock volume is:

$$\Delta V = V_1\left[\exp\left(-\int_{\sigma_1}^{\sigma_2} \alpha_{弹} \, d\sigma\right) - 1\right]$$

ΔV volume increment after rock relaxation;
V_1 rock volume before relaxation;
σ_1, σ_2 effective stress from over burden before and after relaxation;
α_{el} rock elastic compression coefficient.

The rock matrix volume can be regarded as constant during the relaxation process, therefore, volume relaxation increment largely reflects pore volume increment. However, as elastic compression coefficient α_{el} is relevant to valid overburden pressure, which is a function of valid overburden pressure $\alpha_{el} = f(\sigma_{eff})$, Good linear relationship between stress and strain, which means $d\sigma/d\varepsilon$ is constant, only occur at low stress conditions. No linear relationship occurs when stress is high (without plastic deformation or fracture) and $d\sigma/d\varepsilon$ is no longer constant (Van Der Knapp 1599; Dirk Teeuw 1971). Data from Dirk Teeuw (1971) show:

$$E = \frac{d\sigma}{d\varepsilon} = b(\sigma)^q$$

b and q in the equation are constant, q is about 0.33. This formula shows that the greater is the stress, the smaller is the strain increment by the same stress.

The rock pressure release experiments indicate that coefficient of rock pore volume compressibility increases obviously with decreased overburden valid pressure during the relaxation process. The lower is the valid pressure, the greater is the coefficient of compressibility (Fig. 3.33), which is consistent with Hooke's law (Fig. 3.34).

In order to calculate pore volume variation by rock relaxation, quantitative relationship between elastic compression coefficient and valid pressure need to be established, or using average elastic compression coefficient of relaxation under effective pressure before and after uplift. Using average compression coefficient can simplify the calculation of sandstone relaxation pore volume increment.

Experimental simulations of rock uplifting process can be conducted based on the principle that rock elastic compressibility only depends on effective stress rather than absolute pressure. The uplifting process can be regarded as effective pressure release process from overburden. Effective stress added on the rock can simulate different burial depth. If the variation of rock matrix volume is ignored, the variation of rock volume can be used to estimate the change of pore volume, therefore,

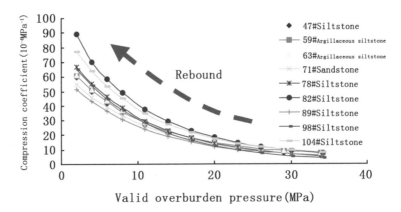

Fig. 3.34 Correlation between elastic compression coefficient and effective over burden pressure in sandstone relaxation experiments

rock relaxation process can be quantified. Relationship between different rock volume and valid overburden pressure or rock volume variation and valid overburden pressure decrement can be established. Figures 3.35 and 3.36 show the experimental results of pore volume changes with overburden pressure and calculated rock volume increment and valid overburden pressure decrement by using a sample from well Yan100 at depth 1715.13 m in the Dongying Sag. The basic information of the sample and experimental results are listed in Tables 3.13 and 3.14. Based on eroded thickness determined in previous section and valid overburden pressure changes during erosion process, the rock volume changes in erosion relaxation process can be quantitatively calculated.

Statistical data from the Dongying Sag indicate that different grain sizes have similar compressibility except for argillaceous sandstone and calcareous sandstone. The elastic compression coefficient depends mainly on the burial depth of the rock. As the primary reservoirs in the Dongying Sag consist of fine silty sandstones, only burial depth was considered in rock relaxation experiments.

The pore volume compression data in the Dongying Sag were classified by depth and the relationship between average rock volume changes and valid burden pressure during the relaxation process of each sample can be established. Statistical results show that sandstones with different grain sizes at the same burial depth have

Fig. 3.35 Relationship between pore volume and effective pressure during sandstone relaxation for a sample from well Yan100 at 1715.13 m

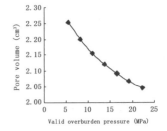

Fig. 3.36 Relationship between pore volume increment and effective pressure decrement during sandstone relaxation for a sample from well Yan100 at 1715.13 m

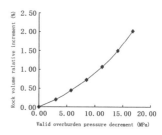

Table 3.14 Results of sandstone relaxation experiments for a sample from well Yan100 at 1715.13 m

Number	Net overburden pressure (MPa)	Pore volume (cm³)	Porosity (%)
1	5.52	2.25	18.94
2	8.27	2.20	18.58
3	11.03	2.16	18.28
4	13.79	2.12	18.03
5	16.55	2.09	17.83
6	19.30	2.07	17.97
7	22.29	2.05	17.50

very similar correlation in rock volume increment and valid pressure decrement; while sandstones at different burial depth show very different relationships between rock volume increment and valid pressure decrement. Under the same decrement of valid pressure, relaxation volume increment in the shallow rocks is much larger than that for a deeply burial one.

3.3.2   Template of Sandstone Relaxation Quantity and Calculation Method

1. The establishment of sandstone relaxation template.

The rock volume increment (relative) has good quadratic function relationship with valid pressure decrement during relaxation from different burial depths (Fig. 3.37). The correlation can be expressed by the following formula and the constant is zero.

$$y = ax^2 + bx$$

y rock volume increment during relaxation;
a, b coefficients controlled by burial depth;
x variation of effective pressure from overburden, MPa.

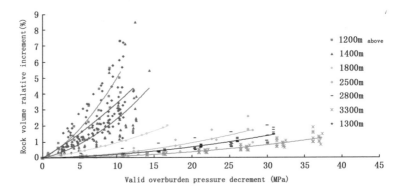

Fig. 3.37 Correlation between pore volume increment and effective pressure decrement during sandstone relaxation in the Dongying Sag

Formulas (3.1)–(3.7) show a detailed calculation of sandstone relaxation for varying burial depth

$$1200\,\text{m} \pm 20\,\text{m} \quad y = 0.281x^2 + 0.217x \quad R^2 = 0.865 \tag{3.1}$$

$$1300\,\text{m} \pm 20\,\text{m} \quad y = 0.008x^2 + 0.2617x \quad R^2 = 0.6097 \tag{3.2}$$

$$1400\,\text{m} \pm 20\,\text{m} \quad y = 0.145x^2 + 0.0977x \quad R^2 = 0.5524 \tag{3.3}$$

$$1800\,\text{m} \pm 20\,\text{m} \quad y = 0.0035x^2 + 0.0592x \quad R^2 = 0.9984 \tag{3.4}$$

$$2500\,\text{m} \pm 20\,\text{m} \quad y = 0.0019x^2 + 0.0082x \quad R^2 = 0.8012 \tag{3.5}$$

$$2800\,\text{m} \pm 20\,\text{m} \quad y = 0.0012x^2 + 0.0093x \quad R^2 = 0.8682 \tag{3.6}$$

$$3300\,\text{m} \pm 20\,\text{m} \quad y = 0.0009x^2 + 0.0007x \quad R^2 = 0.8781 \tag{3.7}$$

A template of sandstone volume increment and valid pressure decrement from overburden can be established based on our dataset (Fig. 3.38). Quantity of sandstone relaxation for samples at other burial depths can be calculated by interpolation.

2. Quantitative computation method of sandstone relaxation.

To directly calculate the sandstone relaxation amount based on a rock elasticity mechanical formula is very difficult because of heterogeneities in sedimentary strata and lack of various actual geological and physical parameters. The relaxation template established by relaxation experimental simulations (Fig. 3.38) is used in

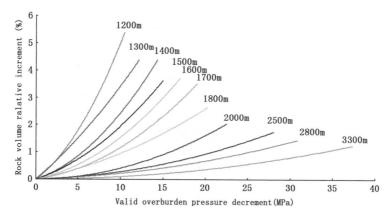

Fig. 3.38 Templates of pore volume increment and effective pressure decrement during sandstone relaxation in the Dongying Sag

the present study to calculate sandstone relaxation. To calculate the sandstone relaxation amount requires reconstruction of the burial history of the studied area. Sandstone burial depth before elevation needs to be determined. The decrement of effective pressure from overburden can be calculated on the basis of eroded thickness and sandstone relaxation amount can be obtained from either the relaxation template or empirical formula. The formula is as below:

$$\Delta V_i = V_i y_i$$

ΔV_i rock volume increment during relaxation;
V_i rock volume before relaxation;
y_i function of rock volume increment;
a, b coefficients of relaxation controlled by burial depth before uplifting;
x decrement of effective pressure.

3.3.3   Application of Sandstone Relaxation Calculation in the Dongying Sag

The template of the sandstone relaxation volume increment and decrement of effective pressure from overburden in Fig. 3.38 and methods described in the previous sections can be applied to calculate sandstone relaxation during the main erosion period in the Dongying Sag. As the unconformity after deposition of the Dongying Formation is the most important one, sandstone relaxation caused pore

Fig. 3.39 Isopach of sandstone thickness increment after relaxation due to erosion at 14.66 Ma in the Dongying Sag

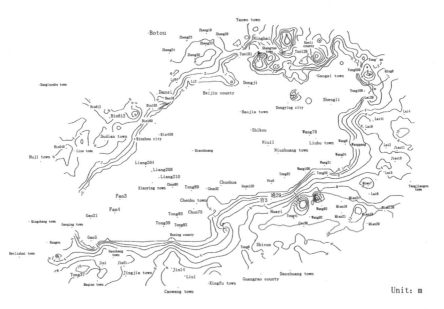

Fig. 3.40 Isopach of sandstone thickness increment after relaxation due to erosion at 11.52 Ma in the Dongying Sag

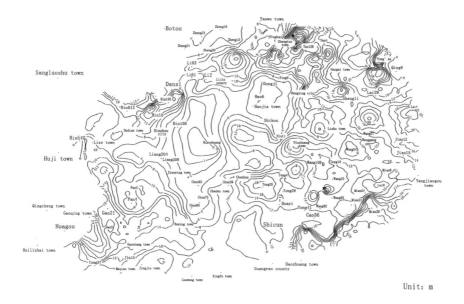

Unit: m

Fig. 3.41 Summed sandstone thickness increment (m) after relaxation in the Dongying Sag

volume increment and effective pressure decrement calculation focus on this unconformity. The calculated results for the Dongying Formation to upper Es4 member are illustrated in map view to show their distributions (Figs. 3.37, 3.38, 3.39, 3.40 and 3.41). Due to variable tectonic settings at different locations, relaxation lasts for different lengths of time in the Dongying Sag, which can be illustrated by comparison Fig. 3.39 with Fig. 3.40. Sandstone relaxation in the central region of the sag such as Qiaozhuang to Haojiazhen area ended at 14.66 Ma, which is much earlier than other regions. Sandstone relaxation ended at 11.52 Ma in the marginal area. The amount of sandstone relaxation also differs significantly at different locations as illustrated in Fig. 3.40. The largest relaxation occurs at southeast and northeast margin of the sag, where sandstone thickness increment is about 20–30 m. Sandstone thickness increment is 3–8 m in Shengtuo, Dongxin and Niuzhuang areas, while there is almost no relaxation in the central sag near Qiaozhuang area.

The unconformity after deposition of the Minghuazhen Formation is much less important than the previous one. The eroded thickness and effective pressure show much less dramatic variations. The thickness of sandstone relaxation in the Dongying Formation is about 2–4 m, which is evenly distributed across the whole sag (Fig. 3.42).

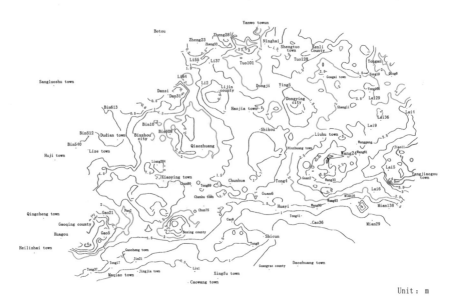

Fig. 3.42 Isopach of sandstone thickness increment after relaxation due to erosion of the Minghuazhen Formation in the Dongying Sag

The sandstone relaxation caused by uplift and erosion and decompression after deposition of the Dongying Formation is about 5–20 times stronger than that of late Minghuazhen Formation. It is not only caused by different eroded thickness, but is also related to burial depth of the Dongying Formation and the older strata. The burial depth of the Dongying and Shahejie formations after deposition of the Dongying Formation is about 1000–1200 m shallower than that after deposition of the Minghuazhen Formation. Sandstone at the shallow burial depth has a larger relaxation energy than a deeply burial one.

The relaxation amount of sandstone is related to the degree of sandstone development, burial depth before elevation and eroded thickness. The relaxation volume increment is positively correlated with sandstone proportion in the strata and eroded thickness, but negatively correlated with burial depth before elevation. The largest relaxation amount occurs in the area where thick sandstone is well developed, shallowly buried and with a high degree of erosion of overburden, such as the southern and northern margins of the sag. On other hand, sandstone relaxation is minimal in the central depression area because there is a low proportion of sandstone development, large burial depth and low erosion.

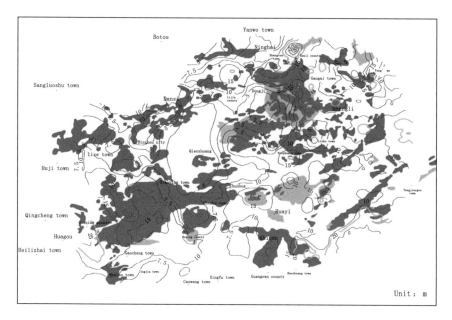

Fig. 3.43 Sandstone thickness increment (m) after relaxation of the Dongying unconformity and distribution of oil fields in the Dongying Sag

The sandstone relaxation amount and distribution map of known oilfields indicates that the high oil abundance such as at Shengtuo, Dongxin and Yong'an oilfields is largely coincident with a high sandstone relaxation degree (Fig. 3.43). As oil accumulation requires pore space and sandstone relaxation can meet such a requirement, the high sandstone relaxation area is favorable for oil accumulation.

Chapter 4
Applications of Basin Formation, Hydrocarbon Generation, and Accumulation Theory

In traditional petroleum geology theory, sedimentary basins are regarded as principal geological units as long as hydrocarbon generation, migration, and accumulation have taken place, regardless of their size, morphology, and genetic type. Therefore, all petroliferous basins are featured with the same or similar petroleum geological evolution processes and characteristics. Based on statistical data and comprehensive analysis of 98 Meso-Cenozoic petroliferous basins (or depressions), common features of hydrocarbon and reservoir formation can be summarized although these basins differ greatly in genetic type, size, and morphological characteristics. All the basins have experienced three stages of development, namely continuous subsidence, overall uplift, and shrinkage. The stage of continuous subsidence is concurrent with the main oil generation from source rocks. By the end of this stage, a great amount of internal energy has accumulated within basins due to the continuous loading of sediments. The overall uplift stage is the time that basins get eroded severely. Petroleum accumulation largely occurs in this stage coincident with pressure decrease and energy release. The shrinkage stage refers to the time period of basin adjustment. Petroleum accumulations get settled. Following this way of thinking, petroleum geological characteristics of each stage can be easily tackled, the inter-relationships with basin formation, hydrocarbon generation, and accumulation could be deeply understood and all kinds of petroleum geological problems can be solved.

This chapter mainly focuses on an introduction of the research and analysis of hydrocarbon generation and accumulation issues in petroliferous basins (or depressions), including data collection, analysis and testing, map compilation, and their application. For better understanding and application, Dongying Sag was used as an example.

© Petroleum Industry Press and Springer Science+Business Media Singapore 2017
D. Guan et al., *Theory and Practice of Hydrocarbon Generation within Space-Limited Source Rocks*, Springer Geology, DOI 10.1007/978-981-10-2407-8_4

4.1   Contents and Approaches of Basin Formation, Hydrocarbon Generation, and Accumulation

4.1.1   Contents of Basin Formation, Hydrocarbon Generation and Accumulation

The theory of basin formation, hydrocarbon generation, and accumulation mainly involves four main aspects: geological evolution history of basins, the history of deposition, and hydrocarbon generation during the stage of continuous subsidence; pressure regime characteristics, and hydrocarbon accumulation history during basin uplift phase; tectonic activities of basins, and hydrocarbon formation history in the shrinkage stage. Details are explained as follows:

1. The development history of basin petroleum geology

In order to understand the characteristics of regional tectonic and development history of geology comprehensively, the geophysical data, stratigraphic formation data, and logging data were collected. These data mainly include: the structural features of the basin base, tectonic unit division, distribution characteristics of the depression area, the uplift region and slope belt, the stratigraphic sequence, sedimentary facies, depositional system characteristics of the main sedimentary sequences, characteristics of the geothermal field and magmatism events, subsidence history, development history of basinal tectonics, and characteristics of tectonic movement. Then these data were comprehensively analyzed and the whole basin evolution processes can be divided into three stages: the continuous subsidence, overall uplift, and shrinkage. Finally, the research results mainly are illustrated in the form of regional graphs and key data, which include multiple layer maps of basin tectonics, main seismic section images of a regional cross section of connecting wells, stratigraphic sections, and isopach graphs and sedimentary facies maps of major target intervals.

2. Diagenesis and hydrocarbon generation during the continuous subsidence stage

There are two essential aspects in the research of diagenesis and hydrocarbon generation during the continuous subsidence stage. One is petroleum geological evolution of the source rocks and the other is the thermal evolution of organic matter. The first part basically involves plots of basic graphs, for example, a thickness isopach map of source rocks and overburden strata, sedimentary facies maps of the main strata, isopach map of sandstone percentage in the main strata, top and bottom age map of the main strata, porosity and depth correlations in both sandstone and mudstone, past and present pressure graphs of the main evolution stage, geographical location and well location maps of the basin (or depression), and boundary location map of the source kitchens. The approach of organic matter thermal evolution assessment derives from thermal simulation experiments. Organic geochemical parameters can be obtained through experimental simulations of hydrocarbon generation and expulsion under variable temperature and pressure regimes, analysis of source rock TOC, extractable organic matter, Ro, homogenization temperature of fluid inclusion, kerogen types, and FAMM. Through comprehensive analysis of source rock diagenesis and kerogen thermal evolution, the overall process and characteristics of hydrocarbon generation during continuous subsidence stage could be obtained.

3. Characteristics of the pressure regime and hydrocarbon accumulation history in the basin uplift stage

In this part, our research mainly focuses on characteristics of the pressure regime before overall uplift (the end of continuous subsidence). Specifically, it includes an investigation of the differential pressure between depression where source rocks are located and structural highs and slopes where reservoir rocks are situated, and correlations between differential pressure and hydrocarbon accumulations. The graph package contains a burial depth map and Ro isopach map of main source rock top and bottom boundary, isopach map of source rock thickness, paleo-pressure regime of source rock occurrence region, planar and cross section map of paleo-pressure regime, primary migration route, and map of sandstone relaxation amount after basin overall uplift by the end of erosion. For compiling these fundamental maps, the erosion amount in the basin should be computed by the approach of Milankovitch cycles or other justified theory, combined with sandstone relaxation amount and parameters obtained in simulation experiments.

4. Vertical tectonic dynamics and hydrocarbon generation and accumulation during the basin shrinkage stage

Basin shrinkage is featured by small amplitude vertical tectonic fluctuation. Each subsidence is accompanied by pressure increase due to sediment loading, and a consequent increase in source rock burial depth, and immature to mature transformation of source rocks which have not reached the threshold of burial depth. Therefore, thermal evolution-related graphs of each subsidence during basin shrinkage should be complied in the same way as those during the late continuous subsidence stage. After a small amplitude of final uplift, the basin shrinkage enters into the Quaternary era which is also called the modern development stage. Therefore, besides the above mentioned graphs, maps of the major exploration target tectonics and graphs of exploration status of the current basin should be complied. Finally, sweetspots for exploration can be pinpointed after comprehensive analysis of all these graphs.

For a better understanding of the application of basin formation, hydrocarbon generation, and accumulation, Dongying Sag was used as an example in Sect. 4.2.

4.1.2   Research Approaches of Basin Formation, Hydrocarbon Generation, and Accumulation

1. Division of petroleum geology evolution stages in a sedimentary basin

According to the research outcomes of petroleum geology development history, the basin can be divided into three development stages, i.e., continuous subsidence, overall uplift, and shrinkage.

A basin petroleum geology evolution stage can be determined by comparison of regional tectonic sections, subsidence and deposition models, stratigraphic development and distribution, sedimentary thickness, stratigraphic age, deposition rate, geothermal gradient, source rock developed intervals, the time scale of uplift and erosion, erosion rate, and eroded thickness with the characteristics of each stage (subsiding, uplifting, shrinkage). Table 4.1 shows the brief petroleum geology histories of Miyang, Dongying, Dongpu Depressions in the Cenozoic. The basins experienced continuous subsidence and deposition stage from Eocene to Oligocene, and hydrocarbons are mainly generated during this time period. The uplift stage is from Oligocene to Neogene, which is characterized by oil migration and accumulation. The reservoir finalization stage occurs after the Neogene.

2. Methods and steps of petroleum generation research

The ultimate purpose of petroleum generation research is to quantify the amount of hydrocarbon generation and expulsion. Parameters for the computation could be obtained from simulation experiments on burial history and thermal evolution history. The steps and methods are specified as follows:

(1) Compile the correlation graphs between porosity and burial depth of effective source rocks

Specifically, collect or determine the porosity of source rock samples at different burial depths and establish a calibrated porosity versus depth profile (Fig. 4.1).

(2) Compile the correlation graphs between effective source rock porosity and maturity

Establish the calibrated depth profiles of effective source rock porosity and maturity (see Fig. 2.16), analyze the relationship between porosity and maturity and tabulate key points (Table 4.2).

(3) Effective source rock maturity calibration

Calibrate the measured vitrinite reflectance value by FAMM analysis.

(4) Determination of oil saturation

Obtain oil saturation data directly from drilling, and then apply the data in laboratory thermal simulation experiments.

Table 4.1 Geological evolution history of Cenozoic Dongying, Dongpu, and Miyang Depressions

System	Series	Strata in Dongying and Dongpu depressions	Strata in Miyang depression	Geological processes and physical field properties	Characteristics of petroleum geology	Characteristics of depression development
Quaternary	Holocene	Pingyuan Fm	Pingyuan Fm	Overall shrinkage, adjustment	Finalized hydrocarbon generation and accumulation	Overall shrinkage
Neogene	Pliocene	Minghuazhen Fm	Fenghuangzhen Fm			
	Miocene	Guantao Fm		Uplifted, unloading	Hydrocarbon accumulation	Uplifting
Paleogene	Oligocene	Dongying Fm	Liaozhuang Fm	Continuous subsidence, loading	Hydrocarbon generation	Continuous subsiding
	Eocene-Paleocene	Shahejie Fm	Hetaoyuan Fm			
		Kongdian Fm	Dacangfang Fm			
			Yuhuangding Fm			

Fig. 4.1 Porosity and depth correlations of source rocks in Miyang Depression

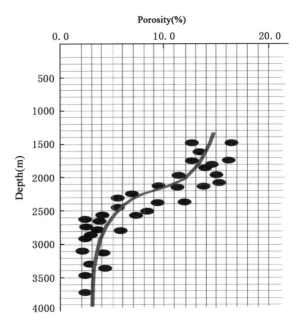

Table 4.2 Corresponding relationships between source rock maturity (%Ro), porosity, and depth

Ro (after calibration) (%)	Porosity (%)	Depth (m)
0.5	17.0	1500
0.6	11.5	2200
0.7	8.5	2700
0.8	7.5	3100
0.9	6.0	3400
1.0	5.5	3750
1.1	5.2	4000
1.2	5	4200

(5) Computation of hydrocarbon generation and expulsion quantity

Select proper parameters for the computation formula in Sect. 4.2.2 and apply it in the computation.

For a better understanding, Dongying Sag is taken as an example.

3. Methods and steps of hydrocarbon accumulation research

① Compile the paleo-pressure maps of each unit in Dongying Sag (including upper Es4 member, lower, middle, and upper Es3 member, Es1 member and Dongying Formation). Figure 4.2 shows the paleo-pressure map of Es2 by the end of deposition of the Dongying Formation.

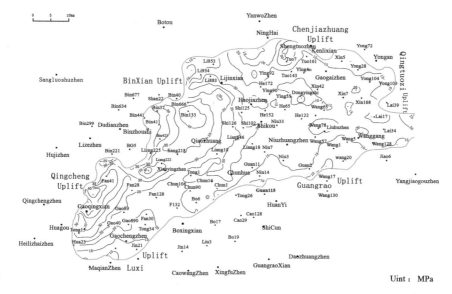

Fig. 4.2 Paleo-pressure map of Es2 member by the end of deposition of the Dongying Formation

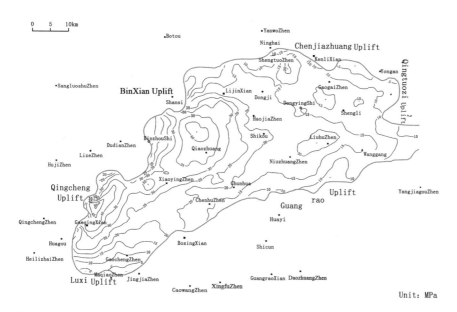

Fig. 4.3 Paleo-pressure map of Es2 member after erosion of the Dongying Formation

② Compile the paleo-pressure maps after erosion of the Dongying Formation
(or before the deposition of the Guantao Formation). Figure 4.3 is the
paleo-pressure map of Es2 member.

③ Compile the isopach map of eroded thickness of the Dongying Sag during Dongying denudation period.

④ Compile the isopach map of the amount of sandstone relaxation of the Dongying Formation in Dongying Sag during the erosion period.

⑤ Compile paleo-burial depth isopach map of each unit from Es4 member to Dongying Formation after denudation (or before deposition of the Guantao Formation). Figure 4.4 is a paleo-burial depth map of Es2 member at the end of deposition of the Dongying Formation.

⑥ Compile isopach maps of sandstone percentage in each stratigraphic unit from Es4 member to Dongying Formation in the Dongying Sag. Figure 4.5 is the isopach map of sandstone percentage of the Es2 member.

⑦ Compile the isopach map of Ro value of the main source rocks by the end of deposition of the Dongying Formation in the Dongying Sag.

By compiling the above maps, the hydrocarbon accumulation process in response to sediment unloading caused pressure release by the end of deposition of the Dongying Formation in Dongying Sag can be elucidated. The favorable sites for hydrocarbon accumulation can be located. By comparison of discovered oil fields, mechanism of petroleum accumulation can be clarified.

The same methods and steps can be applied to other time periods. For example, analysis of sediment unloading and pressure release by the end of deposition of Minghuazhen Formation facilitates the understanding of the present hydrocarbon accumulations.

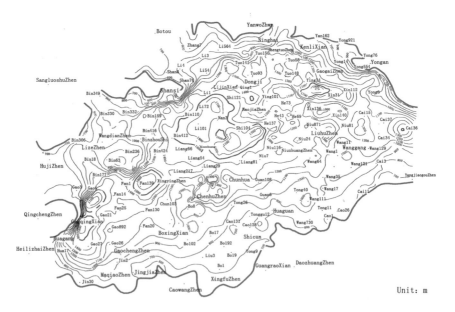

Fig. 4.4 Paleo-burial depth map of Es2 member by the end of deposition of the Dongying Formation

Fig. 4.5   Isopach map of sandstone percentage of the Es2 member

4.2   Quantitative Study of Basin Formation, Hydrocarbon Generation and Accumulation in Dongying Sag

4.2.1   Geological Evolution Stages of Dongying Sag

Dongying Sag is located in the southeast of the Jiyang Depression, with Qingtuozi uplift to the east, Luxi and Guangrao uplifts to the south, Huimin uplift to the west and Chenjiazhuang—Binxian uplift to the north. It is about 90 km long from east to west and 65 km wide from north to south with an area about 5850 km². It is a substructure unit of Jiyang Depression within Bohai Bay Basin in east China (Fig. 4.6) and is a Mesozoic-Cenozoic rift depression created by tectonic evolution in the geological setting of the Paleozoic basement morphology.

Dongying Sag has developed strata of the Paleogene, Neogene, and Quaternary. The Paleogene succession consists of the Kongdian Formation, Shahejie Formation, and Dongying Formation from the bottom upwards, the Neogene succession consists of the Guantao Formation and Minghuazhen Formation, and the Quaternary succession consists of the Pingyuan Formation. The Shahejie Formation is divided into Es1, Es2, Es3, and Es4 members. The deposition of the Shahejie Formation occurs at a major rifting and subsidence phase of this sag with a subsidence rate above 200 m/Ma. The deep depression area is both a depocenter and subsidence center. The depositional environments are deep lake, lakeshore, shallow lake, fan delta, delta, and fluvial. A very thick source rock series was deposited during this

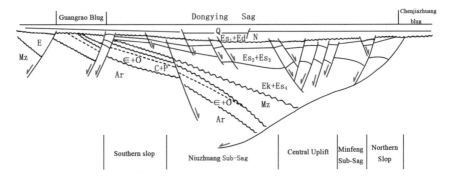

Fig. 4.6 North-south regional geological cross section of the Dongying Sag (adapted from Li Peilong 2000)

stage with the thickest shale sediment more than 3000 m and good source rocks more than 1200 m. Organic materials occur in great abundance with TOC mainly in the range of 1.0–4.0 %. Organic types are dominated by type I and II. The geothermal gradient is relatively high, approximately 40 °C/km. The depression was uplifted as a whole and experienced varying degrees of erosion by the end of the Paleogene due to strong effects of the Himalayan orogeny. The erosion thicknesses obtained using Milankovitch cycles increase from the depocenter where only part of the Dongying Formation was eroded to the depression margin where more strata were removed with total erosion thickness of over 1000 m during 23.03–11.6 Ma. Another cycle but much smaller scale started from the Neogene with 1200–1500 m of sediments being deposited.

According to characteristics of regional tectonic evolution and sequence stratigraphy distribution of the Jiyang Basin, the evolution history of the Dongying Sag can clearly be divided into three stages: continuous subsidence, overall uplift, and shrinkage (Table 4.3).

4.2.2   Research on Hydrocarbon Generation in the Dongying Sag and Parameter Selection for Computation

Most parameters for computation can be obtained directly by exploration, drilling, well logging, and testing, while source rock maturity and porosity evolution can be obtained through experimental simulations on burial history and thermal history of the basin.

1. Computation procedures of hydrocarbon generation

Basin burial history is reconstructed by means of a back-stripping method using parameters such as present burial depth of formations, porosity–depth correlation

Table 4.3 Sedimentation characteristics of Cenozoic Dongying Sag and its petroleum geological evolution

System	Series	Formation	Member	Code	Depth (m)	Geological processes and physical field properties	Characteristics of petroleum geology	Depression development stage
Quaternary	Holocene	Pingyuan Fm		Qp	250–350	Deposition of Guantao and Minghuazhen Formations, fluctuation (energy adjustment)	Finalized	Overall shrinkage
Neogene	Pleistocene	Minghuazhen Fm		Nm	100–1200		hydrocarbon generation and accumulation	
	Miocene	Guantao Fm		Ng	300–400			
					Erosion	End of Dongying Fm deposition, uplift, erosion, unloading, decompression, sandstone relaxation (energy release)	Hydrocarbon accumulation	Overall uplifting
Paleogene	Oligocene	Dongying Fm		Ed	100–800	Deposition of Shahejie Formation, sediment loading, pressure increment, deep burial and heated (energy accumulation)	Hydrocarbon generation	Continuous subsidence
		Shahejie Fm	1st member	Es1	0–450			
			2nd member	Es2	0–350			
			3rd member	Es3	700–1200			
	Eocene		4th member	Es4	<1500			
		Kongdian Fm	1st member	Ek1				
			2nd member	Ek2				

curves of sandstone and mudstone, and the ratio of sandstone to mudstone. The burial depth and porosity of sandstone and mudstone of each formation in geological time have been obtained based on reconstructed thermal history. The computation of thermal history is based on changes in the value of geothermal heat flow over time. The evolution of paleo-temperature was calculated with a heat conduction equation, meanwhile, the time-temperature index (TTI) was calculated. The maturity of source rocks at any burial depth in its geological history can be calculated on the basis of correlations between Ro and TTI. Geothermal history, together with a porosity-depth curve, can be applied to determine the porosity corresponding to source rock mature stage and to compute the theoretical value of maximum oil generation. Oil migration from a relatively high pressure zone to a low pressure one is triggered by differential pressure between source rocks and reservoir rocks caused by erosion. Sediment loading intensifies the compaction effects and causes a further decrease of porosity. The actual amount of oil expulsion can be obtained by calculation of the decrement of porosity at different maturity levels.

The back-stripping technique for burial history reconstruction is based on the principle of mass conservation. With increasing burial depth, sediments are tightly compacted, resulting in a decrease in porosity and formation thickness but intact mass of the rock matrix. The basic principle of back-stripping is an assumption of no change in formation skeleton mass (except for faults). The method is to strip the formations one by one in terms of geochronology in the present stratigraphy, then compute the actual burial depth of each stratum in geological time.

2. Principle of data selection

The investigated strata in the Dongying Sag include the Pingyuan Formation, Minghuazhen Formation, Guantao Formation, Dongying Formation, Es1 member, Es2 member, upper, middle, and lower part of Es3 member and upper Es4 member. Principles of data selection involve the following aspects:

i. Burial history

Strata for experimental simulation: the same as strata for computation.

Absolute age of each stratum: The absolute age of each stratum for experimental simulation is relative to the geochronological boundaries of Cenozoic strata of the Dongying Sag (Table 3.7).

Burial depth of each stratum: stratification data were collected from more than 2800 exploration wells, 10 regional seismic profiles and well correlation maps, structure maps (in terms of burial depth) of strata on different scales. After careful scrutiny, 2333 exploration wells with relatively complete sequence stratigraphy and consistent formation tops, 5 regional seismic profiles with good geological interpretation (The NS-trending 591.7, 618.2 and the WE-trending 92.3, 109, 104.6 lines), 6 cross sections of connected wells (total 453.7 km), and 4 burial depth maps were adapted for regional investigation. Based on that, we plotted a burial depth map of each stratum.

Fig. 4.7 Porosity-depth correlations curve of sandstones in the Dongying Sag

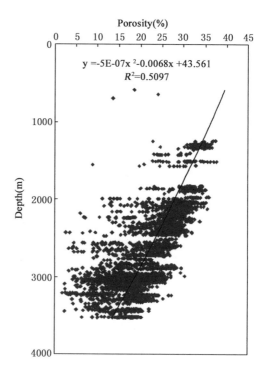

Porosity-depth correlation curves: collect porosity-depth data and profiles, including both sandstones and mudstones. Because those data play a key role in the computation of hydrocarbon generation and expulsion amount, the measured porosity of cores in the Dongying Sag were collected as supplementary material to narrow down error caused by erosion. For this reason, a sandstone porosity-depth correlation curve from the deep sag with relatively small erosion (Fig. 4.7) was chosen. The actual measured data of sandstone porosity from the main oil fields of the Dongying Sag were incorporated into the final determination of the sandstone porosity–depth curve (Fig. 4.8). The mudstone porosity–depth curve is based on the testing data and with reference to surface porosity of standard mudstone (Fig. 4.9).

Lithological distribution and sandstone/mudstone ratio of each stratum: on the basis of depositional facies and sandstone percentage of each target layer available in the Dongying Sag, we plotted isopachs of sandstone percentage based on data from over 500 exploration wells.

Erosion: the main eroded strata in the Dongying Sag occur by the end of Dongying Formation and Minghuazhen Formation. Data of 185 exploration wells were used herein for the identification of Milankovitch cycles recorded in the strata. Through thorough correlation and calibration, a preferred sediment thickness corresponding to 0.405 Ma was restored. Therefore, erosion time after the deposition of the Dongying Formation can be estimated (Fig. 4.10) and erosion thickness in the Dongying Sag can be computed (Fig. 4.11). Data from 57 wells were selected to compute the erosion time (Fig. 4.12) and thickness of the Minghuazhen Formation (Fig. 4.13).

Fig. 4.8 Porosity-depth correlations curve of sandstones

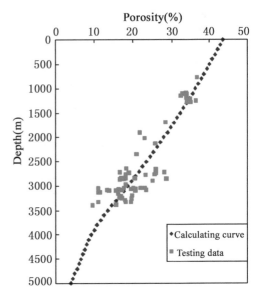

Fig. 4.9 Mudstone porosity-depth profile used for calculations in the Dongying Sag

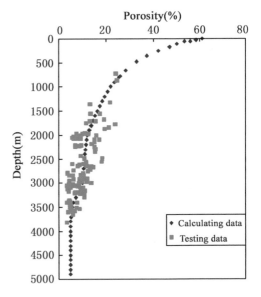

The result shows that Dongying Sag has undergone uplift and erosion since 23.03 Ma. The most intensive erosion occurs in the southeastern part of the basin with eroded thickness of 1000–1700 m. The second largest erosion was in the northeast with eroded thickness of 600–1000 m. Small erosion occurs in the southwest and northwest with eroded thickness of 500–800 m. In the whole depression, the smallest eroded thickness is only several meters, while the largest

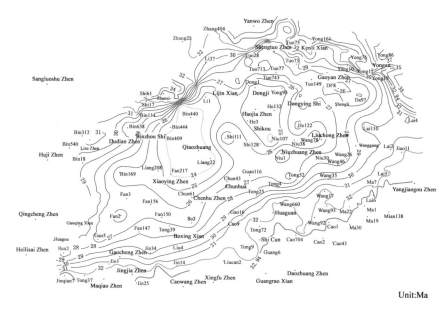

Fig. 4.10 Absolute stratigraphic age of residual strata in the erosion stage of Dongying Formation in Dongying Sag

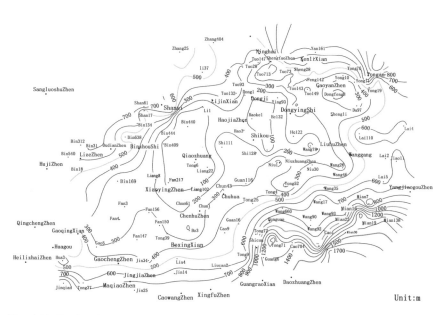

Fig. 4.11 Erosion thickness of erosion stage of Dongying Formation in Dongying Sag

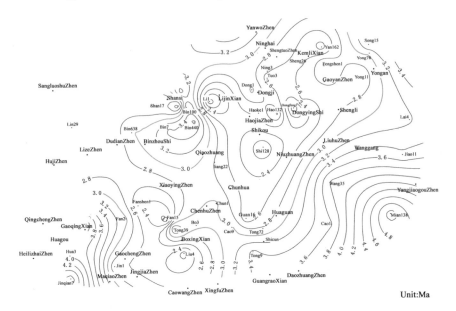

Fig. 4.12 Absolute stratigraphic age of residual strata in the erosion stage of Minghuazhen Formation in Dongying Sag

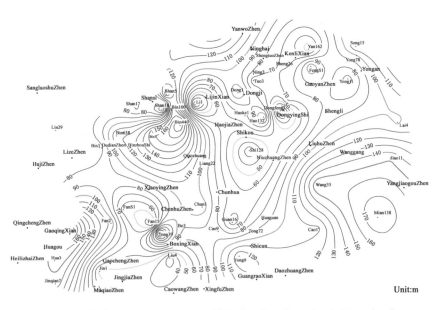

Fig. 4.13 Eroded thickness at erosion stage of Minghuazhen Formation in Dongying Sag

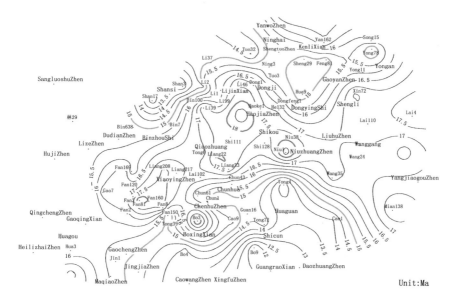

Fig. 4.14 Erosion end timing at denudation of the Dongying Formation in Dongying Sag

thickness is above 1600 m. Erosion of the Minghuazhen Formation is much more mild and uniform with the maximum eroded thickness around 180 m.

Generally either deposition or erosion occurs in nature. In the case of Dongying Sag, the beginning of deposition of the Guantao Formation was regarded as the end of erosion of the Dongying Formation. The time of the end of erosion of the Dongying Formation can be estimated (Fig. 4.14). Early ending occurs in the deep depression where erosion finished around 18.5 Ma, while erosion lasts to 11.5 Ma at the basin margin.

ii. Thermal history

The selection of surface temperature: based on current annual average temperature and paleo-climate conditions since the Paleogene in Dongying area, we selected 15 °C as an average constant temperature near water surface.

Table 4.4 shows the thermal conductivity data for the various strata.

Table 4.5 shows density and specific heat capacity of water and rock.

Ro-TTI correlation curves: regression curves of Ro-TTI based on a correlation of 402 samples from 31 basins by Waples (1980) were used (Fig. 4.15).

iii. Thermal parameter testing

Paleo-temperature obtained by experimental simulations can be tested and calibrated by temperatures measured in drilling wells. In light of the perfect consistency between simulated present temperature and the well testing temperatures (Fig. 4.16), the validity of simulated paleo-temperature in this method is convincing.

Table 4.4 Thermal conductivities of the studied strata (Gong Yuling 2003)

Horizon	Es4s	Es3x	Es3z	Es3s	Es2x	Es2s	Es1	Ed	Ng	Nm	Qp
Thermal conductivity (W/mK)	1.98	1.81	1.81	1.81	1.7	1.7	1.9	2.09	1.97	2.04	2.04

Table 4.5 Density and specific heat capacity parameters (Yukler 1978; Han Yuji 1986)

Lithology	Parameter	
	Density ρ (g/cm^3)	Specific heat capacity (Cal/g °C)
Sandstone	2.65	0.197
Mudstone	2.68	0.223
Water	1.004	1.008

Fig. 4.15 Correlation curve of Ro and TTI (Waples 1980)

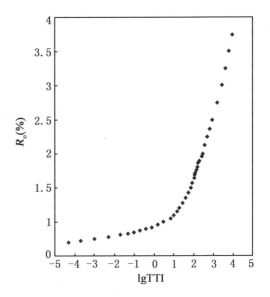

The measured Ro value was used to examine the validity of the simulation experiments on source rock maturity. The simulated Ro values match well with measured ones at different burial depths, which justify the outcome of thermal simulations (Fig. 4.17).

4.2.3   Characteristics of Main Source Rocks at Different Evolution Stages and Quantification of Generated Oil

1. Characteristics of main source rock evolution by the end of the Dongying Formation deposition and quantity of oil generated

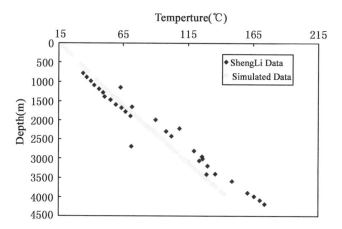

Fig. 4.16 Comparison of simulated and measured values of present temperature at various burial depths in Dongying Sag

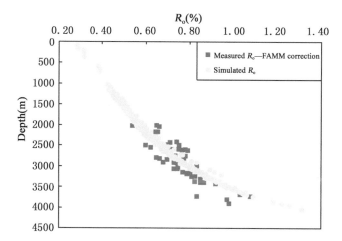

Fig. 4.17 Comparison between simulated and measured values of present Ro at various burial depths in Dongying Sag

There are four hydrocarbon kitchens in the Dongying Sag—Lijin, Niuzhuang, Boxing, and Minfeng sub-sags. The source rocks exist throughout the middle and lower part of Es3 and the upper part of Es4 members. Lijin sub-sag is buried most deeply, followed by Niuzhuang, Minfeng, and Boxing sub-sags. By the end of deposition of the Dongying Formation, the paleo-geothermal gradient was 38.8 °C/km. The source rocks of the middle Es3 member were not deeply buried and only a small portion is mature. The shallowest burial depth was 600 m in southern part of the depression, while the deepest depth was 2500 m in Lijin sub-sag (Fig. 4.18), where Ro was in the range of 0.45–0.6 % (Fig. 4.19).

Fig. 4.18 Burial depth of the bottom boundary of middle Es3 member in late deposition stage of Dongying Formation in Dongying Sag

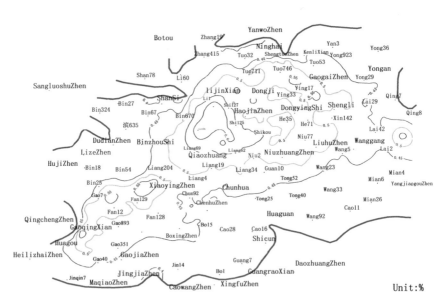

Fig. 4.19 Ro values of middle Es3 member in late deposition stage of Dongying Formation in Dongying Sag

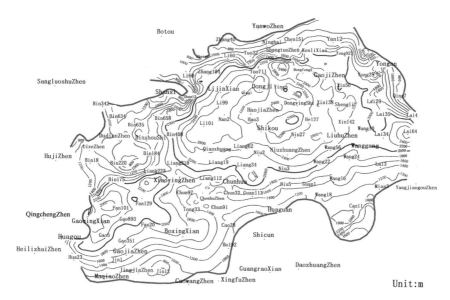

Fig. 4.20 Burial depth of the bottom boundary of lower Es3 member in late deposition stage of Dongying Formation in Dongying Sag

The quantity of oil generated from middle Es3 in the Dongying Sag can be estimated according to the depression evolution stages.

By the end of deposition of the Dongying Formation, the source rocks of the lower Es3 member has partially been buried up to 2800 m (Fig. 4.20), which is beyond oil generation threshold (Fig. 4.21). The quantity of oil generated from the lower Es3 member can be computed accordingly.

The source rocks of the upper Es4 member were buried deeply and largely mature, in the oil generation window (Figs. 4.22 and 4.23). Therefore, we can compute generated oil quantity in the upper Es4.

2. Characteristics of source rock evolution by the end of deposition of the Minghuazhen Formation and quantity of oil generated

By the end of deposition of the Minghuazhen Formation, the basin has been in the shrinkage stage, with the paleo-geothermal gradient decreased from 38.8 °C/km by the end of deposition of the Dongying Formation to 35.5 °C/km, and burial depth increased 1000–1200 m. This evolution process has rendered most source rock enter the oil window. Some parts of middle Es3 were deeply buried (Figs. 4.24 and 4.25). The quantity of oil generated from the middle Es3 member can be calculated.

By the end of deposition of the Minghuazhen Formation, some of the source rocks of the lower Es3 have been deeply buried (Fig. 4.26), beyond the oil generation threshold (Fig. 4.27). The quantity of oil generated from the lower Es3 member can be calculated.

Fig. 4.21 Ro value of lower Es3 member in late deposition stage of Dongying Formation in Dongying Sag

Fig. 4.22 Burial depth of the bottom boundary of upper Es4 member in late deposition stage of Dongying Formation in Dongying Sag

Fig. 4.23 Ro value of upper Es4 member in late deposition stage of Dongying Formation in Dongying Sag

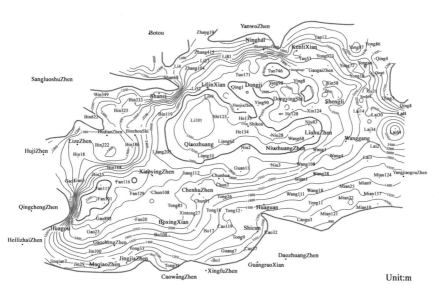

Fig. 4.24 Burial depth of the bottom boundary of middle Es3 member in late deposition stage of Minghuazhen Formation in Dongying Sag

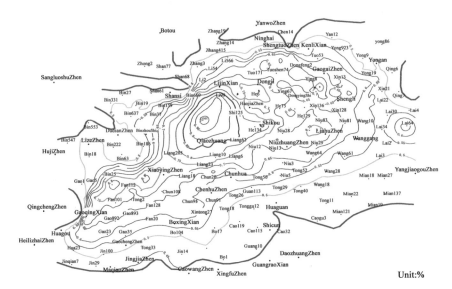

Fig. 4.25 Ro value of middle Es3 member in late deposition stage of Minghuazhen Formation in Dongying Sag

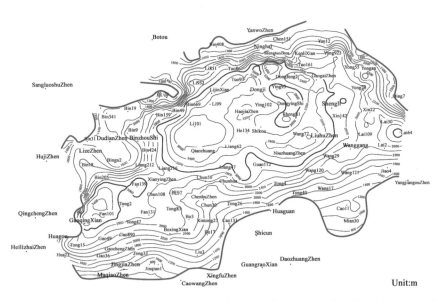

Fig. 4.26 Burial depth of the bottom boundary of lower Es3 member in late deposition stage of Minghuazhen Formation in Dongying Sag

By the end of deposition of the Minghuazhen Formation, source rocks of the upper Es4 have been deeply buried (Fig. 4.28) and largely matured (Fig. 4.29). The generated oil from the upper Es4 was quantified.

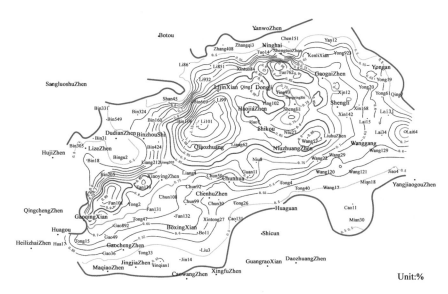

Fig. 4.27 Ro value of lower Es3 member in late deposition stage of Minghuazhen Formation in Dongying Sag

4.2.4   Analysis of Oil Expulsion

1. Oil expulsion quantity by the end of deposition of the Dongying Formation

By the end of deposition of the Dongying Formation, Lijin sub-sag has the highest oil expulsion rate among four source kitchens. About 48 % of the total expulsion was expelled at this time, followed by Niuzhuang sub-sag, which accounts for 27 % of total expulsion, while Minfeng and Boxing sub-sags have low expulsion rates of 13 and 12 %, respectively (Fig. 4.30).

From the above analysis, it can be concluded that basin overall uplift stage is the main period for oil expulsion, accounting for 48 % of total expulsion over the whole basin.

2. Oil expulsion during the deposition of the Guantao-Minghuazhen formations

The deposition progressed evenly and slowly throughout the Guantao-Minghuazhen period. The pressure regime was inherited from the erosion period of the Dongying Formation. The overlying strata of source rocks were increased by approximately 1000–1200 m in this period. The increased compaction of source rocks results in oil migration to the lower pressure regime of reservoir rocks formed during erosion period of the Dongying Formation (Figs. 4.31, 4.32 and 4.33).

Fig. 4.28 Burial depth of the bottom boundary of upper Es4 member in late deposition stage of Minghuazhen Formation in Dongying Sag

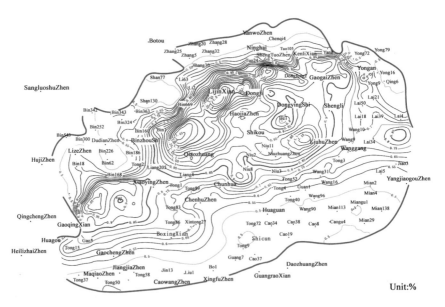

Fig. 4.29 Ro value of upper Es4 member in late deposition stage of Minghuazhen Formation in Dongying Sag

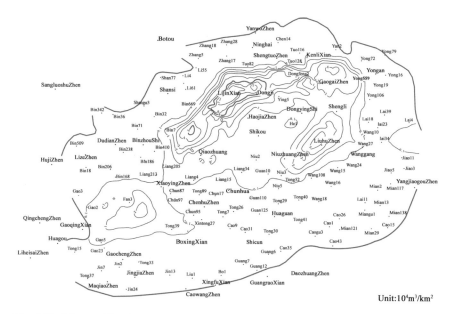

Fig. 4.30 Oil expulsion intensity of upper Es4 member in late erosion stage of Dongying Formation in Dongying Sag

Fig. 4.31 Oil expulsion intensity of middle Es3 member in late erosion stage of Minghuazhen Formation in Dongying Sag

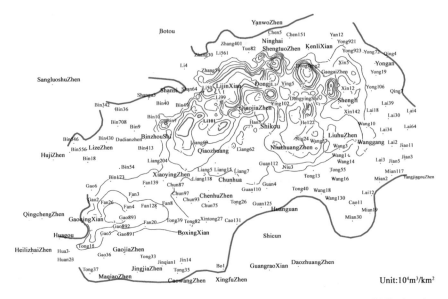

Fig. 4.32 Oil expulsion intensity of lower Es3 member in late erosion stage of Minghuazhen Formation in Dongying Sag

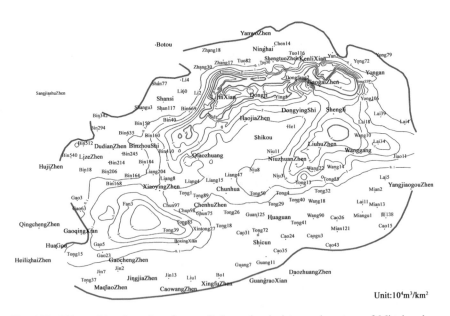

Fig. 4.33 Oil expulsion intensity of upper Es4 member in late erosion stage of Minghuazhen Formation in Dongying Sag

The Niuzhuang sub-sag has the highest oil expulsion rate among the four sags in this period, contributing 35 % of total oil expulsion, followed by Lijin sub-sag with 31 %, while Boxing and Mingfeng sub-sags have low expulsion rate of 19 and 15 %, respectively.

Data illustrated here suggest that the deposition of Guantao-Minghuazhen formations was another critical time period for oil expulsion, accounting for 47 % of total expulsion in the whole basin.

Chapter 5
Oil Generation from Marine and Lacustrine Environments

There has been a debate for a long time between western and Asian academics about whether oil can be found in freshwater sediments. After decades of exploration, Chinese petroleum geologists not only found oils in the Mesozoic–Cenozoic lacustrine sedimentary basins, but also established several giant oil fields with over hundred million tons of reserve, such as Daqing and Shengli. The achievements justified their persistence on the perception of lacustrine oil generation. However, there are still some issues that have not been solved yet such as the differences between marine and lacustrine originated oils, mechanisms and theories of marine and lacustrine oil generation even though it is widely accepted that both marine and lacustrine sediments can generate oil. In this chapter, preliminary analysis and discussion about these scientific issues will be addressed.

5.1 Comparison Between Marine and Lacustrine Source Rocks

5.1.1 Data Selection for a Comparative Study

Statistically, more than 80 % of the world's oil and 90 % of China's oil resources have been found in the Jurassic, Cretaceous, Paleogene, and Neogene strata. While most of these stratigraphic intervals are marine sediments worldwide, all of them are lacustrine deposits in China. Comparison of four lacustrine source rocks developed in China with representative marine source rocks of the same age from all over the world facilitates understanding of the similarity and differences between them. To make the results more informative, both geography and basin type are considered. The Zagros Basin in the Middle East represents a foreland basin, which has the biggest conventional oil reserves. The Los Angeles Basin in America represents an active continental margin basin which has the highest abundance of

© Petroleum Industry Press and Springer Science+Business Media Singapore 2017
D. Guan et al., *Theory and Practice of Hydrocarbon Generation within Space-Limited Source Rocks*, Springer Geology, DOI 10.1007/978-981-10-2407-8_5

oil and gas. Campos Basin in Brazil represents a passive margin basin which has the greatest petroleum potential. Songliao Basin and Bohai Bay Basin represent lacustrine rift basins which have the most abundant oil resources in China.

5.1.2 Depositional Environment and Characteristics of Marine Source Rocks

1. The original idea that oil can be generated only from marine source rocks

In 1859, Edwin Drake from Pennsylvania was in charge of drilling the first petroleum exploration well at Oil Creek, in a village called Titusville. Oil was encountered at a depth of 21.2 m in August with production of 4.1 tons/day after pumping. Oils were generated from the Paleozoic marine black shale. Following this well, a large number of oil wells in the Oil Creek area were drilled. Oil production reached 410×10^3 tons/day by 1862 in this area. Meanwhile, a large number of petroleum wells appeared in western of Pennsylvania State, southwestern New York State and Ohio State. All the oil under production was generated from the Paleozoic black marine shale. This gave birth to the explosive boom of oil production of the Appalachian Basin, which reached 5×10^6 tons/day by 1900. All these achievements made the Appalachian Basin become the center for petroleum production from the Paleozoic strata. The great abundance of petroleum resources discovered in the Paleozoic marine source rocks from the Appalachian Basin drew the attention of many geologists. Bacteria, phytoplankton, zooplankton, and higher plants were found to be the main sources of organic matter in the sediments by a comparative study of petroleum of marine origin and coal. As higher plants appeared in the Middle Devonian, phytoplankton in the ocean was regarded as the only source of organic matter for petroleum generation in the early Paleozoic. Meanwhile, phytoplankton in oceans was considered as the initial source of all the organic matter in sediments, which accounts for the origin of natural derivatives of aquatic plants in the ocean. Coal is the natural derivative of terrigenous plants and coal-related organic matter carried to oceans by rivers occupies less than 1 % of the total organic matter in the ocean. The quality as well as the quantity of the oceanic organic matter is much better than that from land. Phytoplankton and zooplankton in the ocean are rich in protein and lipids, which are the main source for marine oil generation. On the contrary, terrigenous higher plants have lower protein and lipid contents but higher cellulose and lignin contents and their contribution to oil generation is limited.

In conclusion, Western scholars represented by American scholars considered that "lacustrine source rock cannot generate oil" mainly based on two reasons. The early petroleum exploration showed that the upper Paleozoic oils in the United States all from marine strata. Marine planktonic algae contain oil-prone organic matter, while such organic materials are absent in lacustrine environments.

2. The description of oil marine origin in *"Petroleum Formation and Occurrence"*

From the 1930s to the 1940s, disciplines like petroleum geology and geochemistry quickly emerged, which promoted research into the marine origin of oil and marine source rock as well as breakthroughs on such topics. For example, conditions of organic material formation were largely derived from the perspective of petroleum geology, organic origin of petroleum, and marine origin of oil. Sources of organic matter and their yields in the oceanic environment and chemical composition are based on life evolution in the biosphere. Many discoveries about the marine origin of oil and marine source rocks had been made by the 1960s and 1970s. Tissot and Welte (1978) published *"Petroleum Formation and Occurrence,"* based on numerous research outcomes. They first put forward the theory of kerogen thermal degradation for hydrocarbon generation.

Their book emphasizes the perspective of marine origin of oil. The formation and accumulation of organic matter have been illustrated from the perspective of geology in the first part of the book. The main content and vital points are summarized here.

First, the yield of organic matter in oceanic environment is very important for the formation of marine source rocks on the basis of the analysis of photosynthesis and the evolution of the biosphere. Unicellular, microorganisms and phytoplankton are the main producers of the organic matter in oceans. Diatoms, dinoflagellates, blue-green algae, flagella, immobile cells, and micro plankton are the major groups. The organic material is mainly produced by phytoplankton, which formed the primary link of the biological chain, and thus became the precursor of life. Diatoms, dinoflagellates, and coccoliths are the three most important producers in the phytoplankton group. The second link of the biological chain is zooplankton (such as small crustaceans), which feed on the phytoplankton.

Second, bacteria, phytoplankton, zooplankton (mainly foraminifera and crustaceans), and higher plants are main suppliers of the organic matter in the sediments. All organisms basically have the same chemical components, such as lipids, proteins, carbohydrates, and lignin in higher plants. For the generation of oil, lipid is the most important component, which includes fatty substances, wax, and pseudo-lipid components, such as oil-soluble pigment, terpane compounds, steroid, and more complex fats. Therefore, bacteria, phytoplankton, and zooplankton are the main contributors of organic matter to marine source rock. Terrigenous higher plant material is the fourth major source of organic matter in sediments, most of which contains 50–70 % cellulose and lignin with extremely small proportions of lipid and protein. Thus, the organic matter coming from terrigenous plants is less likely to generate oil and more likely to form coal.

Third, the fact that abundant oil was found from the Cambrian, Ordovician, and Silurian strata in the United States and other countries proves the marine origin of oil. Since plants had not emerged and become widely distributed on land until the Devonian, marine phytoplankton organisms, bacteria, and blue-green algae are the main suppliers of oil in the Paleozoic. Two periods of obvious high yields of marine phytoplankton worldwide can be found in Fig. 5.1. The first started in the

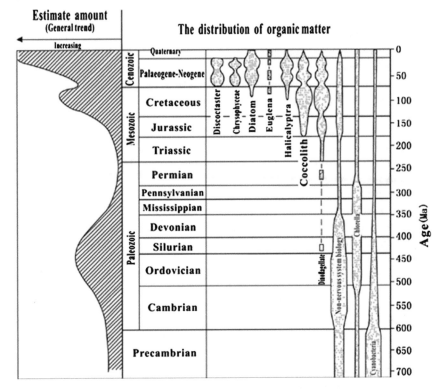

Fig. 5.1 Variations in phytoplankton abundance in different geological times (After Tissot and Welte 1978)

Precambrian and reached a peak in the Paleozoic, corresponding to this peak was abundant oil of marine origin worldwide in lower Paleozoic strata. Especially, the discovery and rapid growth of marine origin oils in upper Paleozoic in America made it the birthplace of marine oil. The second period is late Jurassic–Cretaceous–Paleogene–Neogene. Corresponding to this peak was a large number of oil fields found in the Mesozoic–Cenozoic worldwide.

3. Characteristics of marine source rocks in the United States and elsewhere

The distribution of petroleum in America is directly controlled by the regional geological conditions. The Paleozoic basins in central and northern regions were developed on a stable platform and oils were produced from the Paleozoic strata. Back-arc basins were developed in the west and south of the platform and oils were produced from Paleozoic and Mesozoic strata. The thrust fault zones were developed around the platform area and oils were produced from Paleozoic and Mesozoic strata. The coastal plain and the shelf region have produced oil from Cenozoic strata. Intermontane basins in southern California have produced oils from Cenozoic strata. The North Slope basins in Alaska have produced oils from

Cenozoic strata. Almost all oils are produced from marine strata. Although lacustrine source rocks do occur, such as in Utah and Wyoming Paleogene lacustrine shale, their contributions to petroleum accumulations are trivial. The organic matter of early Paleozoic marine source rocks is mainly from bacteria and blue-green algae. Mesozoic oils are mainly from coccoliths and dinoflagellates, and Cenozoic groups are mainly from flagellates and diatoms.

Zagros Basin in the Middle East has four marine source rock sequences: Cambrian, Silurian, Jurassic–Cretaceous, and Paleocene–Eocene. All organic matter of these source rocks originated from marine algae with type II kerogen. Jurassic–Cretaceous marine shale is the main source rock. The marine source rock in Los Angeles Basin is mainly Miocene diatom shale which contains abundant algae with an average TOC content around 5 %. Kerogen types are a mixture of Type I and II, which contain 85 % sapropel maceral and has great potential for hydrocarbon generation. The main source rock of Campos Basin in Brazil is the Lower Cretaceous Lagoa Feia lacustrine black calcareous shale, which is a non-marine sequence formed during rift development. Although the original organic matter was formed in a lacustrine environment, organic matter is basically composed of simple algae dominated by type II kerogen formed under strongly reducing and high salinity conditions with TOC up to 5 %.

In summary, marine source rocks were developed globally under saline water (salinity 10–35 ‰) and brackish water (salinity 1–10 ‰). The organic matter is largely from marine phytoplankton such as blue-green algae, dinoflagellates, coccoliths and diatoms. Good marine source rocks generally have TOC contents of 3–5 % and are dominated by type II kerogen. They are mostly black shales or black calcareous shales with thickness of 100–300 m.

5.1.3 Proposal of Lacustrine Origin of Oil

In the 1920s and 1930s, most petroleum was found in marine strata all over the world. This fact had convinced many geologists in the United States and Western countries of the idea that only marine source rocks are capable of oil generation. In 1913, the petroleum companies of the United States and other western countries had designated groups of geologists to carry out a geological survey in China. They almost gave up further investigation and drilling before they made any breakthrough. Some of them asserted lacustrine source rocks have no oil generation potential. China oil-producing formations or oil seepage in China were marine sourced and Mesozoic and Cenozoic marine strata were not deposited in China. Therefore, they concluded that China was poor in oil.

Since the 1930s, some Chinese geologists gradually performed geological surveys in the north of Shaanxi, the Hexi Corridor, Sichuan and Tianshan region. Dushanzi oil field in Paleogene lacustrine strata in Xinjiang and Laojunmiao oil field in Cretaceous lacustrine strata in Yumen of Gansu were found in 1938 and 1939 successively. The most noteworthy is that professor Zhongxiang Pan, the first petroleum geologist in our country, who, after graduating from Peking University in 1931, made four surveys in the north of Shaanxi, and many surveys in Sichuan and other places. He pointed out that oil in Shaanxi originated from Triassic and Jurassic lacustrine strata. There are many favorable structures for petroleum accumulations in Sichuan and large oil fields may be encountered if oil accumulation occurs. During Dr. Pan's Ph.D. in the university of Kansas in the United States, he found supporting examples for his argument in the literature that the oil field in northwest of Colorado was produced from Paleogene lacustrine sediments. So he presented his thesis titled "Non-marine origin of petroleum in North Shensi, and the Cretaceous of Szechuan, China" at the AAPG annual conference in 1941, putting forward the lacustrine oil generation theory in China for the first time.

The lacustrine oil generation theory greatly inspired Chinese petroleum geologists to explore Chinese lacustrine basins, released them from the fetters of the assertion that there is no oil generated from lacustrine sediments in China. After decades of exploration, a large number of giant lacustrine oil fields have been discovered, with oil production exceeding 10^8 tons in 1976. These successful explorations have not only proved the great abundance of oil in Chinese lacustrine basins, but also justified the great potential of petroleum generation in lacustrine basins. Moreover, it has provided a large number of practical data and experience for geologists to study characteristics of petroleum geology in lacustrine systems. The lacustrine oil generation theory has widely been accepted by petroleum geologists all over the world.

5.1.4 Developments in Lacustrine Oil Generation in China

There are several stages in the research history of Chinese lacustrine oil generation theory.

1. Stage focus on petroleum geology

In the 1930s–1950s, exploration wells in lacustrine sedimentary basins in China were rare and the study of lacustrine oil generation was mainly based on outcrop observations. The organic carbon contents and depositional environments of source rocks are largely based on outcrop measurement and sample analysis. For example, the petroleum geology investigation report of Xinjiang pointed out that all lacustrine dark mudstones of the Upper Permian, Triassic, Jurassic, and Paleogene might be source rocks, especially the black shale and oil shale of the Upper Permian, which are organic rich and have favorable conditions for oil generation. The Laojunmiao oil field was discovered in 1939. Some geologists proposed that oils were derived from the Cretaceous deep-water lacustrine black shales, which have numerous fossils and low fixed carbon ratio (55–61 %). Since the middle of the 1950s, with the increase of drilling in sedimentary basins, petroleum investigation and exploration work gradually shifted from outcrops to the subsurface. However, the study of lacustrine oil generation theory in the subsurface of basins is quite different from geological surveys at outcrops. The petroleum geological field survey is often linear, which means to observe the change of formation, perpendicular to bedding strike. However, in the coverage area of basins, investigation largely relies on the data from drilling, logging, testing and seismic gravity, and magnetic surveys. Here we use Songliao Basin as an illustration to review the research of lacustrine oil generation in 1950s–1960s.

(1) Research approaches

The deep depression of the Songliao Basin was located starting with basin evolution processes. The subsiding center and depocenter were compared to identify whether they are the same or not. The depositional environment was investigated to identify sediment formed under highly reducing conditions. Organic geochemistry and reservoir parameters were derived from sample analysis. The assemblages of source–reservoir-cap were identified. On the basis of the relationship between source rocks and reservoir rocks, potential exploration targets for oil and gas were located.

(2) Research contents

The research contents include basin evolution history and structural unit division; the development and sedimentary characteristics of the depression area; the depositional environments of the depression area (including salinity of ancient lacustrine basins, pH value, redox index, paleo-climate, and water changes of lacustrine basins); geological and geochemical characteristics of source rocks (including lithology, thickness, organic carbon content, and degree of organic matter transformation); source rock rank evaluation (Table 5.1); source rock assessment for each group; computation of generated oil quantity; classification and evaluation of source–reservoir-cap rock assemblage; source kitchen development characteristics; reservoir distributions, and reservoir locations.

(3) Source rock evaluation and quantity of oil generated from each group in the Songliao Basin

Through a comprehensive analysis of the abundance of organic matter in the source rocks, the redox conditions of the depositional environment, the transformation ratios of the organic matter, and the sedimentary conditions of the source rocks, it was eventually found that the first member of Qingshankou Formation is the main source rock which has the most favorable conditions for oil generation and potential. The second and third members of the Qingshankou Formation and the first member of the Nenjiang Formation are important source rocks, and the second and third members of Yaojia Formation are source rocks which have fairly good potential.

According to the basic understanding that extractable organic matter (chloroform bitumen "A") is oil residual in source rocks, which has very similar properties to produced oil, a computation formula for quantity of oil generation is as follows:

$$Q = V \times \rho \times A_1 \times A_{oil} \times K$$

Q Oil generation;
V Volume of source rock;

Table 5.1 The criteria for lower Cretaceous source rocks in Songliao Basin (from internal report of the Songliao Basin 1977)

Level of source rocks	Organic matter						Sedimentary environments			Transformational conditions			Geologic features			Tectonics
	C (%)	Fluorescence and dispersed asphalt					S (%)	K	Geochemical facies	L_1	L_2	L_3	Lithologic properties	Sedimentary cycles	Sedimentary facies	
		B (%)	Oil	Pectin and Asphaltene	A_1 (%)	A_1/A_2										
Fine	>0.5	>0.015	>5.0	<85	>0.05	>0.8	>0.4	>0.4	Reduction—Strong Reduction	>7	>20	>30	Dark gray mudstone and black shale	Mainly steady sinking	Deep and semi-deep lacustrine	Depressions
Fair	0.5–0.2	0.015–0.005	5.0–2.0	85–95	0.03–0.05	0.6–0.8	0.15–0.4	0.25–0.4	Weak Reduction—Reduction	5–7	15–20	20–30	Mainly gray mudstone, partly gray green mudstone	Crustal starts to return	Shallow lacustrine	Transitions
Poor	<0.2	<0.005	<2.0	>95	<0.03	<0.6	<0.15	<0.25	Oxidation	<5	<15	<20	Purple, red, brown and gray green mudstone	Relatively rise or vibration frequently	Shore lacustrine, fluvial and partly shallow lacustrine	Uplift

Note L_2 is the second bituminization coefficient, which is the ratio of the percentage of carbon in the total bitumen to the percentage of surplus organic carbon in the rock. It reflects the capacity of organic matter transformation into bitumen

L_3 is the third bituminization coefficient, which is the ratio of the percentage of carbon in chloroform bitumen "A" relative to the carbon in the total bitumen. It reflects the richness of neutral free bitumen in total bitumen

ρ   Density of source rock;
A_1 The content of chloroform bitumen "A" in source rock;
A_{oil} The content of oil components in chloroform bitumen "A";
K The coefficient of oil components transformed into oil, 0.4 is used for the
       lacustrine source rock of the Songliao Basin.

The calculation results of the oil generation in each group are

1st Member of the Qingshankou Formation: 4.8×10^9 t;
The second and third member of the Qingshankou Formation: 1.6×10^9 t;
1st Member of the Nenjiang Formation: 0.8×10^9 t;
The second and third member of the Yaojia Formation: 0.03×10^9 t;
The total quantity of generated oils in the Songliao Basin: 7.3×10^9 t.

The basic understanding of lacustrine oil generation derived from this stage is that lacustrine source rocks not only can generate oils but can form giant oil fields. Organic rich, thick source rocks are formed over the long-term in a stable subsidence region with humid and semi-humid climate conditions. Humid climate favors a large quantity of organic matter accumulation; the reducing depositional environment in freshwater lakes facilitates organic matter preservation and the long-term subsidence is favorable for oil generation.

2. The stage of organic geochemistry study

Since the middle 1970s, numerous advanced scientific theories and technology have been brought into China due to economic reform and international exchange policies. Kerogen thermal evolution for oil generation, experimental methods, and technologies exerted an extensive impact on the research of petroleum geology and the study of lacustrine source rocks in China. Because of the introduction of the theory and experimental methods, petroleum geology research shifted to organic geochemistry research, and these geologists became pioneer organic geochemists to apply the theory and technologies.

From the 1980s until now, achievements of marine and lacustrine source rocks are all on the basis of this theory and experimental methods. It should be pointed out that this theory is derived from global marine source rocks and should be applicable in marine origin oils. However, organic geochemists in China have applied this theory and experimental techniques to study specific issues of lacustrine origin oil without scrutiny. They termed this as the theory of lacustrine origin of oil. For better understanding, several examples are illustrated here.

(1) The hydrocarbon generation patterns from marine and lacustrine organic
      matter

Figures 5.2 and 5.3 show the comparison patterns between Chinese hydrocarbon generation from lacustrine source rocks with that of marine source rocks. Essentially, no difference can be observed.

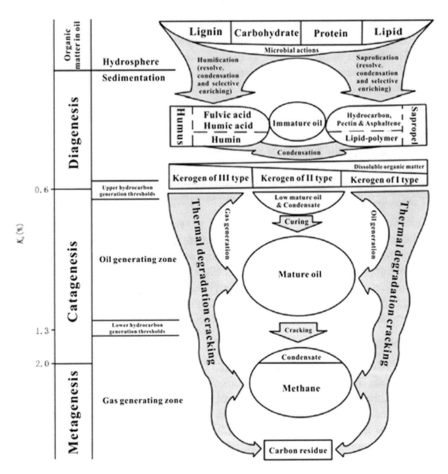

Fig. 5.2 Evolutional model of organic matter in lacustrine source rocks, China (50 years of Chinese petroleum science and technology 2000)

(2) The evolution characteristics of hydrocarbon generation from Chinese lacustrine source rocks

It can be seen from Fig. 5.4 that the natural evolution characteristics of source rocks in Mesozoic and Cenozoic strata are consistent with those proposed by Tissot and Welte (1978) in terms of source rock thermal evolution and hydrocarbon generation.

(3) Chinese lacustrine source rock assessment criteria

Tables 5.2, 5.3, and 5.4 show examples of assessment criteria of Chinese lacustrine source rocks and organic matter abundance.

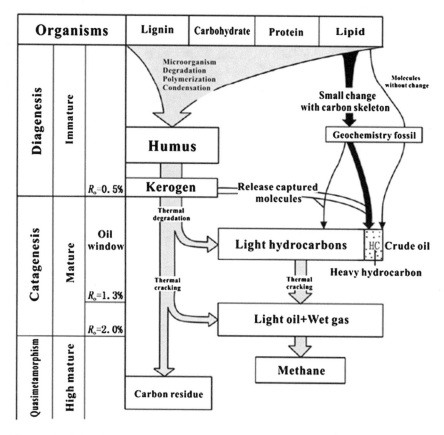

Fig. 5.3 Evolutional model of organic matters in marine source rocks (Tissot and Welte 1978)

3. The stage to study depositional environments and biological characteristics of lacustrine systems

Since the late 1980s, the Science and Technology Development Bureau of China National Petroleum Corporation started to investigate the paleontology of Jurassic, Cretaceous, Paleogene, and Neogene systems in oil and gas provinces of China for a deep understanding of the basic geology of the lacustrine basins. The initial purpose was to locate prospective areas for oil and gas through depositional environments and microfossils in Chinese lacustrine source rocks, especially from the perspectives of paleontology, paleo-environment, and paleo-climate. The research focuses on a comprehensive analysis of biostratigraphy and paleo-environments of the Cretaceous system of the Songliao Basin, the Paleogene and Neogene System of the Bohai Bay Basin, and the Jurassic system in northern China. Some remarkable breakthroughs have been made through more than 10 years of scientific research.

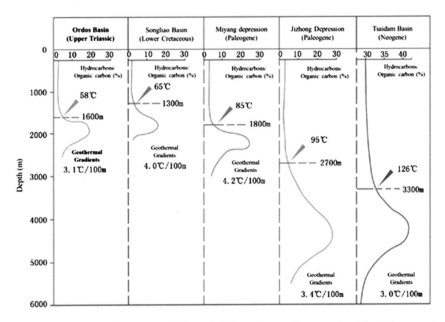

Fig. 5.4 Natural evolution curve of Cenozoic–Mesozoic organic matter in China (50 years of Chinese petroleum sciences and technology 2000)

Table 5.2 Assessment criteria of lacustrine source rocks (Pilong Li 2003)

Items	Good	Fair	Poor	None
Lithofacies	Deep and Semi-deep lacustrine	Semi-deep and shallow lacustrine	Shore lacustrine	Fluvial
Type of kerogen	Sapropelic	Sapropel-Humic	Humic	Humic
H/C ratio	1.7–1.3	1.3–1.0	1.0–0.5	1.0–0.5
TOC (%)	3.5–1.0	1.0–0.6	0.6–0.4	<0.4
Chloroform bitumen "A" (%)	>0.12	0.12–0.06	0.06–0.01	<0.01
Total hydrocarbons (μg/g)	>500	500–250	250–100	<100
Total hydrocarbon/ TOC (%)	>6	6–3	3–1	<1

(1) Depositional environment and biological characteristics of the Songliao Basin

Cretaceous shale from the first member of Qingshakou Formation, and dark shale from the first member of the Nenjiang Formation have thicknesses of 80 and 100 m and TOC of 2.2 and 2.4 %, respectively, which form the major source rocks of the Songliao Basin.

Available geological and geophysical data indicate that Songliao Basin is an inland lacustrine basin. The depositional environments of the Qingshakou and

Table 5.3 Evaluation criteria of organic matter abundance of lacustrine source rocks in China (50 years of Chinese petroleum sciences and technology 2000)

Source rock rank	Good	Fair	Poor	None
Lithofacies	Dark gray and grayish mudstone	Gray mudstone	Gray green mudstone	Red mudstone
TOC (%)	>1.0	1.0–0.6	0.6–0.4	<0.4
Chloroform bitumen "A" (%)	>0.1	0.1–0.05	0.05–0.01	<0.01
Total hydrocarbons (μg/g)	>500	500–200	200–100	<100
S1 + S2 (mg/g)	>6.0	6.0–2.0	2.0–0.5	<0.5
Total hydrocarbon/TOC (%)	20–8	8–3	3–1	<1

Nenjiang formations have been controversial for a long time among domestic and overseas scholars. Mimang Zhang and Jiajian Zhou suggested a marine environment on the basis of *Martostoma-Sunggrichthys* assemblage, which is only discovered in marine strata, while Zhiwei Gu and others claimed a lacustrine environment on the basis of bivalves such as *Striarca, Mytilus,* and *Musculus,* which live in salty and brackish water. These fossils also occur in the Nenjiang Formation. However, most domestic and overseas scholars referring to biocenosis characteristics considered that marine depositional environments or littoral sediments occur in the Songliao Basin.

Paleontologists Dequan Ye and Chuanben Zhao, worked in the Songliao Basin, deem that the salinity of inland lacustrine basins is constantly changing, and that salinity increases in an arid climate, so that the emergence of salty and brackish water species is truly possible.

On the basis of research outcomes of dinoflagellates, Ruiqi Gao suggested that Songliao Basin has experienced two rapid marine transgressions during a long history of lacustrine deposition. The main evidence is as follows:

① There are dinoflagellate assemblages that live in brackish water in the Nenjiang Formation and Qingshakou Formation. *Choate Koransium, Cleistosphaeridium,* and *Sentusidinium* are abundant in the Qingshakou Formation, while *Cleistosphaeridium nenjangensis, Batiacasphaera,* and *Bosedinia* are rich in Nenjiang Formation. Most of these species are from sea water, and their variances emerge in brackish water. Some brackish water species such as *Muscula, Mytilus,* and *Hama* occur as well (Fig. 5.5).

② As is shown in Fig. 5.5, there is a spike in paleo-salinity, a reduction in the alkalinity during the deposition period of the first member of the Qingshankou Formation and the first and second member of the Nenjiang Formation. The water salinity is about 5–10 ‰, belonging to the category of brackish water. With the increasing of heavy isotope values, it approaches a marine environment.

Table 5.4 Average organic matter abundance of Cenozoic and Mesozoic source rocks in main lacustrine rift basins, East China (Pilong Li 2003)

Basin	Songliao Basin		Bohai Bay Basin							Nanxiang Basin	Jianghan Basin
Depression			Jiyang depression			Liaohe depression	Raoyang depression		Dongpu depression (North)	Miyang depression	Qianjiang depression
Stratum	Qing-1	Nen-1	Upper Shahejie-4	Lower Shahejie-3	Shahejie-1	Shahejie-3&4	Shahejie-1	Shahejie-3	Shahejie-3&4	Hetaoyuan-3	Qianjiang
Main type of organic matter	I, II_1	I, II_1	I	I	I, II_1	I, II_1	II_1	II_1	I, II_1	I, II_1	II_2
Organic carbon (%)	2.207	2.402	2.24	2.5	2.58	1.94 ~ 3.36	1.1	0.86	1.01	1.83	0.63
Chloroform asphalt "A" (%)	0.553	0.2804	0.3947	0.3361	0.2957	>0.13	0.1792	0.1742	0.13	0.2168	0.2461
Total hydrocarbon (%)	0.1612	0.1682	0.1785	0.1647	0.1283	>0.05	0.1008	0.0971	0.08	0.1219	0.0969

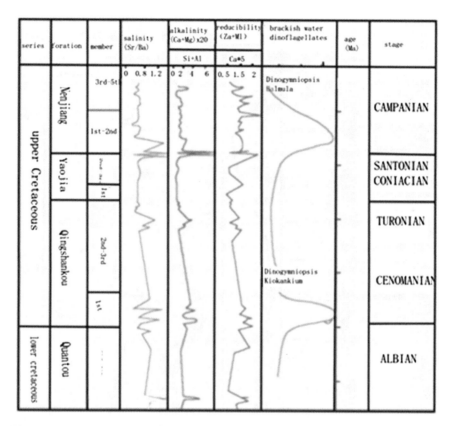

Fig. 5.5 Medium conditions of Cretaceous lacustrine facies and vertical variations in dinoflagellate content (after Ruiqi Gao 1997)

③ During the deposition of the 1st member of the Qingshankou Formation and the first and second member of the Nenjiang Formation, with the subsidence of the basin, the area of the lake expanded rapidly, forming a vast, deep-water environment, bringing the lake closer to the coast (Fig. 5.6). Sea-level rose globally during this period, and brackish water in a lacustrine depositional environment formed by the linking of the lake and ocean.

④ The biomarker distribution characteristics in the source rocks are closely related to the lacustrine brackish water depositional environment and algae. The precursor of tricyclic terpanes is bacteria or algae. Shales of the first member of Nenjiang Formation and the first member of Qingshankou Formation are formed during marine transgression period, which is characterized by high tricyclic terpane content with the ratio of tricyclic terpanes to hopanes of 0.14–0.26. Meanwhile, high gammacerane content indicates strongly reducing environments. Relatively abundant pregnanes and 4-methylsteranes suggest an algal or phycoplast microorganism.

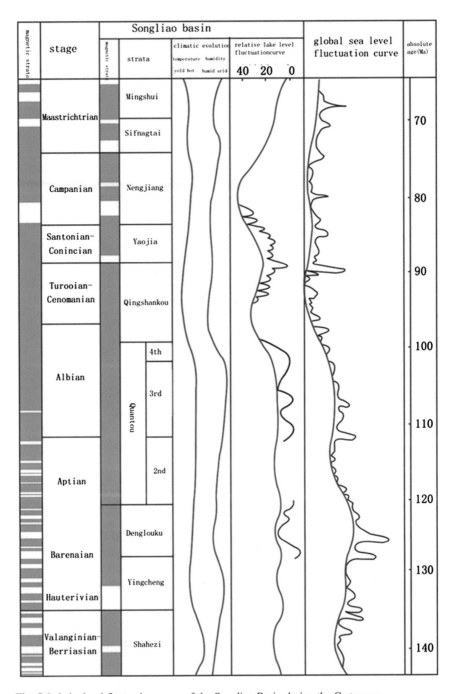

Fig. 5.6 Lake level fluctuation curve of the Songliao Basin during the Cretaceous

⑤ During the deposition period of the first member of the Qingshankou Formation and the first member of Nenjiang Formation, there are abundant dinoflagellates, blue-green algae, and green algae. Algae are rich in protein and lipid content, which are oil prone and can generate oil early. More C_{27} steranes generated from phycoplast were found in low maturity oils suggesting that phycoplasts are the main source supplier for low maturity oil.

In conclusion, the basin has experienced two transgressions, causing temporary links between the lake and the ocean. Algae are the main organic components in the brackish water.

(2) Depositional environment and biomass characteristics of Bohai Bay Basin

TOC contents of the Paleogene Shahejie Formation shale are generally more than 2 %, making it the main source rock of the Bohai Bay Basin. The proved petroleum reserves and production account for more than 90 % of these from the Paleogene and Neogene in China.

There are abundant oil resources in this basin because of the riches of dinoflagellate in this salty water and brackish water during the deposition of the Shahejie Formation. The plentiful dinoflagellates provide a significant amount of organic matter, forming high quality source rocks of the Shahejie Formation. However, some paleontologists believe that dinoflagellates exist only in marine strata that would result from transgression during the deposition of the Shahejie Formation. This perception will certainly be concerned with the issue of whether the source rocks belong to a lacustrine or marine depositional environment. Resolution of this dispute will provide answers to whether the petroleum of Bohai Bay Basin is of marine or lacustrine origin.

To settle the dispute, the China National Petroleum Corporation (CNPC) conducted specific research of "salt lake depositional environments and hydrocarbon generation in the Cenozoic in China." Based on a comprehensive summary of the paleontology, mineralogy, geochemistry, petroleum geology, and other related data of the Cenozoic sedimentary basins in China, marine fossils, special facies-indicating minerals and geochemical characteristics are proposed to exist in inland salty lakes. In terms of organic geochemistry, characteristics of oils and source rocks might differ from typical lacustrine basins and marine basins. Organic matter similar to marine facies could coexist with terrigenous dominant organic matter. In the middle of the inland lacustrine basin evolution, there is a 'High Mountain–Deep Basin' stage when salt can erode from the crust due to deep rooted faults, which have nothing to do with marine transgressions. These results completely changed the traditional understanding of the depositional environment and biomass features, which are either marine or lacustrine. Moreover, the decisive role of a brackish-salty water depositional environment and algae were emphasized in the lacustrine origin of petroleum.

Here the Paleogene and Neogene source rocks of Bohai Bay Basin are used as an example to illustrate the decisive effects on depositional environment of inland salty lacustrine basins and dinoflagellates on the hydrocarbon generation potential of source rocks.

(1) The effect of the planktonic algae on hydrocarbon generation potential of source rocks

There are abundant brackish water and salty water dinoflagellate fossils in the Bohai Bay Basin. For example, during deposition of the Shahejie Formation in the Liaohe rift depression, dinoflagellates account for up to 77 % of the algae in the western sag with a minimum of 66.8 % (average 71.7 %). The maximum proportion is 55.6 % in the eastern sag and the minimum is 39.7 % (average 48.3 %). The Damintun sag has a range of 28.1–45.1 % dinoflagellates (average 36.4 %). It can be clearly concluded through these data that dinoflagellates are rich in the Liaohe rift depression. The analogues can be found in the Jiyang faulted depression and off shore area of the Bohai Bay Basin (Table 5.5).

There is a close relationship between the dinoflagellate enrichment and the oil accumulation in the Dongpu faulted depression. Oil plays are found where there are abundant dinoflagellate fossils. The extensive dinoflagellate fossils are also discovered in the source rocks in other districts, such as the Qianjiang Formation in the Jianghan Basin, the Hetaoyuan Formation in the Biyang fault depression, the Nadu Formation in the Baise Basin, and the Liushagang Formation in the Beibuwan Basin. All high potential source rocks are rich in dinoflagellates in Chinese Tertiary strata (Fig. 5.7).

The variation in the dinoflagellate fossil content also affects the potential of hydrocarbon generation. For example, during deposition of the Dongying Formation in the Liaohe rift depression, the planktonic alga content is low. The planktonic alga contents in the eastern and western sags are 42 and 28.6 %, respectively, while the dinoflagellate contents are low, only 10 and 1 % of that in the Shahejie Formation. Therefore, potential for hydrocarbon generation of the Dongying Formation decreases greatly. The Jinhu Depression in the Subei Basin, a typical shallow freshwater lacustrine basin for example has freshwater alga content of more than 50 % in the Paleocene Funing Formation source rocks, with a maximum content of 80 %. The source rock is humic and sapropelic-humic type, mainly composed of type II2 and III kerogen. The potential for oil generation is low.

(2) The effect of water depth in lacustrine basins on source rock hydrocarbon generation potential

There is a close relationship between deep-water lacustrine basins and commercial oil discovered in the Paleogene and Neogene basins in China. For example, a nearshore-shallow lacustrine facies prevails most of the time and over most of Qaidam Basin with patchy relatively deep salty lacustrine facies on the northwestern rim. Therefore, oil-gas reservoirs are only discovered in the west of the basin (Fig. 5.8).

Table 5.5 Abundance of floating algae in Paleogene of the Liaohe rift depression (Tertiary petroleum provinces in China 1993)

Strata		Western sag		Eastern sag		Damingtun sag	
		Algae/(algae + spore) (%)	Dinoflagellate/algae (%)	Algae/(algae + spore) (%)	Dinoflagellate/algae (%)	Algae/(algae + spore) (%)	Dinoflagellate/algae (%)
Dongying		42	6.1	28.6	0.4		
Shahejie	Es1	21.8	66.8	17.3	55.6	19.6	41.4
	Es3	23.3	77.6	17.3	39.7	8.2	28.1
	Es4	18.1	67.9			5.9	45.1

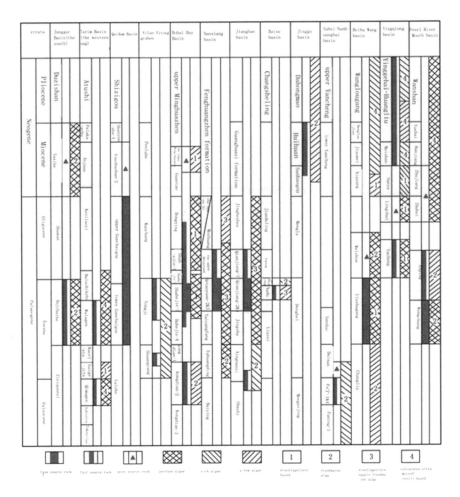

Fig. 5.7 Relationship between the Paleogene and Neogene source rocks and algae fossils (Tertiary oil-gas fields in China 1993)

(3) Effect of water salinity on hydrocarbon generation potential

There are no pure Paleogene and Neogene freshwater lacustrine basins discovered in China. A brief period of freshwater deposition only occurs in the early or late stage of lake development. Brackish or salty water deposits occur in the main period of the lake development, providing favorable conditions for dinoflagellates to proliferate. Water salinity directly affects source rock hydrocarbon generation potential. For example, the upper Es4 to Es3 members of the Shahejie Formation in the Dongpu depression of the Bohai Bay Basin is cut into north and south parts by an E-W trending uplift. The typical salty depositional environment occurs in the north where source rocks are humic and sapropelic-humic types. Types I, II and III kerogens are 21.6, 62.9 and 15.3 %, respectively. However, fresh water

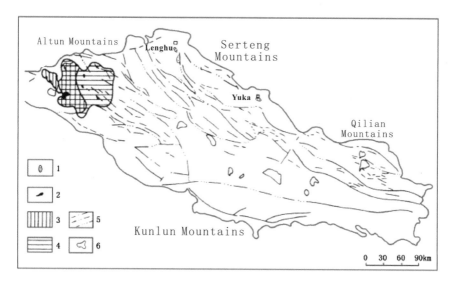

Fig. 5.8 Stacked map of Qaidam Basin Paleogene–Neogene semi-deep, deep lacustrine oil fields (Tertiary oil provinces in China 1993). *1* Jurassic oil fields; *2* Paleogene–Neogene oil fields; *3* Paleogene semi-deep, deep lacustrine oil fields; *4* Miocene semi-deep, deep lacustrine oil fields; *5* Boundaries between structure units; *6* Present lakes

Table 5.6 Statistical of source rock kerogen types in the Dongpu depression (Tertiary oil-gas fields in China 1993)

Region	Strata	Proportion of kerogen type				Sample number
		I	II$_1$	II$_2$	III	
The north	Upper Es4, Es3	21.8	45.2	17.7	15.3	124
The south	Upper Es4, Es3		18	26	56	50

depositional environments prevail in the southern part, where source rocks are dominated by humic type kerogen but type I kerogen is absent. Type III kerogen is up to 56 % (Table 5.6). This accounts for the fact that more than 90 % of the oil reserves are discovered in the north, while only 10 % are discovered only in the south.

In conclusion, Paleogene and Neogene dinoflagellates in brackish to salty water environments in the Bohai Bay Basin are the main component in the source rocks.

(3) Depositional environments and biomass signatures of the Jurassic lacustrine basins in north China

Dinoflagellate fossils have not been discovered in the Jurassic lacustrine basins in north China except for littoral-neritic fossils at the Ussuri Bay in the Sanjiang Plain in the west of Heilongjiang Province. However, there is considerable Botryococcus developing in the lacustrine depositional environment (Table 5.7).

Table 5.7 Algae distribution in the Jurassic lacustrine basins in north China (The Jurassic in north China 2003)

Area	Strata	Botryococcus	Algae	Acritarchs	Lithology
Kuqa depression, northern Tarim Basin	Lower Qigu Fm	Moderate	–	–	Dark mudstone
	Middle Kezilenur Fm	Few	–	–	Dark mudstone
	Upper Kezilenur Fm–Qiakemake Fm	Moderate	–	–	Dark mudstone
	Upper Yangxia Fm	Few	–	A few *Kuqaia*	Dark mudstone
Turpan–Hami Basin	Qiketai Fm	Few	*Schizosporis, Leiaosphaeridia*, a few *Granodiscus*	–	Dark mudstone
	Xishanyao Fm	–	*Schizosporis* (24.5 %), a few *Concentricytes*	–	Coal seam, dark mudstone
	Sangonghe Fm	Moderately high	*Psiloschisosporis granodiscus*	A few *Kuqaia*	Dark mudstone
	Badaowan Fm	Moderately high	–	–	Coal seam, dark mudstone
Qaidam Basin	Seventh Member of Dameigou Fm	Moderately high	*Schizosporis* (average 4.6 %)	–	Tan-beige shale
	Sixth Member of Dameigou Fm	–	*Schizosporis* (12.35 % average)	–	Dark mudstone
	Fifth Member of Dameigou Fm	–	*Granodiscus, Dictyotidium* (3.25 %)		Dark mudstone
	Third Member of Dameigou Fm	–	*Granodiscus, Acritarchs* (71 %)		Dark mudstone
	Second Member of Dameigou Fm	–	*Granodiscus, Acritarchs* (6.31 %)		Dark mudstone
	First Member of Dameigou Fm	–	*Schizosporis* (0–1.73 %)	–	Dark mudstone
	Bottom of first member of Dameigou Fm–Xiaomeigou Fm	–	*Leiosphaeridia, Schizosporis, Granodiscus* (average 14 %)	–	Dark mudstone
Hexi Corridor	Upper Wangjiashan Fm	High	*Schizosporis* (average 12.35 %)	–	Dark mudstone
Ordos Basin	Lower Anding Fm	Moderately high	–	–	Black-brown shale, dark mudstone
	Yanan Fm	Few	*Schizosporis* (average 0.2 %)	–	Black shale
	Fuxian Fm	Few	*Schizosporis* (2.0 % average)	–	Black-brown shale, dark mudstone

(continued)

Table 5.7   (continued)

Area	Strata	Botryococcus	Algae	Acritarchs	Lithology
Erlian Basin	Gerile Fm	Few	–	–	Dark mudstone
Shandong	Fangzi Fm	–	–	A few *Kuqaia*	Dark mudstone

Botryococcus belongs to *Xanthophyta* or *Chlorophyta* and *Botryococcus braunii* is the only species in that category that is widespread in variable salinity water and brackish water. Most strikingly, it has a thick arcus adiposus, appearing pale blue with a grape-like shape under fluorescence microscopy. The arcus adiposus is a product of photosynthesis. Numerous data demonstrate this alga has hydrocarbon content of 30–70 %, and is known as "oil algae." Many references indicate that it is an important component of sapromyxite, lignite, and oil shale. The oil yield of the modern Botryococcus in Darwin, Australia is estimated to be 35 tons per hectare, which equals to 3500 tons per square kilometer. Therefore, not only in the geologic history, but also the modern Botryococcus has high potential for oil generation. High quality source rocks can be found where Botryococcus is discovered.

5.1.5   A Few Summary Markers

According to the comparative analysis of marine and lacustrine source rocks, the following conclusions can be drawn:

1. Both marine and lacustrine source rocks are formed in brackish-salty water depositional environments. Continental shelf, slope, and margin are the best environments for marine source rock formation. Deep-water sags and inland saline basins are the best environments for lacustrine source rock formation.
2. The main source of organic matter of marine and lacustrine source rocks is planktonic alga, which develop in brackish and salty water. In Early Paleozoic, the planktonic alga of marine source rocks is dominated by blue-green algae, while dinoflagellates, diatoms and coccoliths prevailed in the Mesozoic and Cenozoic. The planktonic alga in lacustrine source rocks is primarily dinoflagellata (one class of *Pyrrophyta*) in the Mesozoic and Cenozoic. In the Jurassic, the alga is mainly Botryococcus, because dinoflagellates appeared in inland salty basins in the late Jurassic.
3. Lacustrine and marine source rocks contain type I and type II kerogen respectively, having high potential for oil generation. Formed in inland saline basins, the source rocks with type I kerogen are more favorable for oil generation, and are the best source rocks.

5.2 Issues Concerning Lacustrine and Marine Oil Generation Theories

A comparative analysis revealed no distinct differences between marine and lacustrine petroleum generation, and both of them are formed in the brackish-salty water environments. Living in brackish-salty water depositional environments, planktonic algae are the primary source of organic matter. Actually, the debate on whether oil can be found in freshwater sediments is triggered by the one-sided conclusion drawn through surface appearances in the early stages of petroleum research and exploration. Petroleum geochemistry research achievements and global petroleum exploration practice proved that petroleum either of marine origin or lacustrine origin forms in similar depositional environments, from the same source of organic matter, share the same lithologic features, undergo the same process of sedimentation-diagenesis and kerogen thermal evolution. Therefore, there are no separate theories on marine or lacustrine oil generation or local oil generation theory that is only applicable to a country or a specific region. Since any country or region is just the specific name of an administrative unit in the world, characteristics of petroleum geology in any place of the world are a part of the global petroleum geology evolution progress, thus subject to the general principles of global petroleum geology and possessing singular characteristics. Judging from the perspective of petroleum geology, the only difference is kerogen types formed in three distinct depositional environments.

5.2.1 Depositional Environment of Inland Saline Lacustrine Basins and Characteristics of Kerogen

Statistically, land only covers 29 % of the earth surface, while inland lakes cover only 1 % globally and only 0.8 % in China. The oil kitchens mainly develop in areas with the most subsidence, where dark shale develops. So the ratio of dark shale areas to lake areas is small. For instance, the total area of Songliao Basin is 260×10^3 km^2, and the area of dark shale of the first member of Qingshankou Formation is 62×10^3 km^2, 23 % of the basin area. The distribution area of dark shale of first member of Nenjiang formation is 48×10^3 km^2, only 18 % of the basin area. Hence, compared with marine basins, inland lacustrine basins are fewer with smaller areas, making the lacustrine petroleum exploration harder. However, inland saline lacustrine basins do have advantages over marine basins in terms of conditions for oil generation. For one thing, inland saline lacustrine basins are a restricted environment enclosed by land in all directions, where the oil kitchen has the most subsidence and burial depth, resulting in great water depth and reducing environments. For another, despite of the inflow of floods and streams into a lacustrine basin, they do not have enough energy to stir the bottom water of the oil kitchen, posing no risk to the preservation of organic matter, but which is degraded for a large part by microbes. Thus, kerogen is most often type I, mainly derived from reprocessed organisms and lipids from other microorganisms. Type I kerogen has extremely high H/C ratio (about 1.5 or beyond) and low O/C ratio (generally lower than 0.1), which has the highest potential for producing petroleum. In Mesozoic and Cenozoic inland saline lacustrine basins, petroleum is found to be directly related to deep-water dark shale-rich in dinoflagellates or other algae, developing in the stage of steady subsidence of lake basins, despite the small areas of lacustrine basins and the timing of oil generation. This is the reason why oil kitchens in China's inland lake basins have small areas but are rich in petroleum resources.

5.2.2 Marine Depositional Environments and Characteristics of Kerogen

Ocean covers 79 % of the earth. Continental shelves cover 7.5 % and continental slopes cover 12 % of the ocean, with the rest being the deep sea. For now, petroleum is mainly discovered in continental shelf and continental slope where planktonic algae bloom. Oil generation sags occupy 27 % of the ocean's area. Compared to continents, oil generation sags of the ocean are larger in number, area, and have a wider-range of oil distribution and more oil generated. That is the reason why petroleum was found in marine strata in early exploration, and marine petroleum exploration is easier. However, the marine depositional environment is less favorable in terms of oil generation than the terrigenous depositional environment. Oceans are connected to each other on a global scale, sharing the same sea level, with the oil generation sags (or depressions) in relatively depressed areas in continental shelves and slopes. Be it water depth or quietness, they cannot make the ocean a completely reducing environment. Besides, the general water depth of continental shelves and slopes is tens to hundreds meters, even the maximum depth of 1–2 km of a continental slope bottom is far less than the water depth of deep sags in inland lake basins. This geography and environment is home to terrigenous plant clasts and oxidized deposits, which change the original organic composition. Consequently, kerogen of type II mainly forms in the ocean, which has lower H/C ratio, lower potential for oil generation and more aromatic and naphthenic cyclic compounds. This type of kerogen is capable of generating both oil and gas.

5.2.3 Transitional Depositional Environments and Characteristics of Kerogen

Transitional environments between ocean and land refer to swamp, tidal zone, and littoral-neritic area of the coast. Transitional environments between lake and land refer to swamp and shallow water regions on the lakeshore. The sedimentary organic matter comes from terrigenous higher plants. The organic matter was oxidized debris on the land, and then re-transported and deposited as inertinite and soil humin, forming type III kerogen. This kind of kerogen has extremely low H/C ratio (<1.0) and high O/C ratio (up to 0.2 or 0.3). It consists of polycyclic aromatic nuclei, ketone and carboxylic acid groups composed of heteroatoms, with no ester groups. Source rock with type III kerogen usually has accompanying coal beds in the form of coal seams interbedded with source rocks. Type III kerogen has a low potential for oil generation but strong capacity for generating gas.

References

AAPG (2001) Global proliferous basin series. Petroleum Industry Press, Beijing

Fu C (2000) Chinese petroleum science and technology for 50 Years. Petroleum Industry Press, Beijing

Gao R, Cai X (1997) Oil and gas formation conditions and distribution patterns. Petroleum Industry Press, Beijing

Guan D et al (2004) Theoretical consideration of basin formation, hydrocarbon generation and accumulation—from basin to petroleum reservoir. Petroleum Industry Press, Beijing

Hu W, Chen D (1995) Oil and gas field distribution patterns and exploration experiences in USA. Petroleum Industry Press, Beijing

Li P et al (2003) Petroleum geology of Lacustrine rift basin and exploration. Petroleum Industry Press, Beijing

Liu Z et al (2009) Chinese oil shale. Petroleum Industry Press, Beijing

Sun Z, Yang F et al (1997) Chinese saline lacustrine depositional environment in the Cenozoic era and petroleum generation. Petroleum Industry Press, Beijing

Tissot BP, Welte DH (1978) Petroleum formation and occurrence—a new approach to oil and gas exploration. Springer, Berlin

Xu D (2005) Astro-geologic time scale and the advancement of cyclostratigraphy. J Stratig 29:635–640

Xu D, Yao Y et al (2007) Astrostratigraphic study of the oligocene dongying formation in the Dongying depression, Shandong. J Stratig 31(S2):56–67

Xu D, Yao Y et al (2008) Astrostratigraphic research on the Neogene Minghuazhen Formation in Dongying Sag, Shandong Province. J Palaeogeogr 10:287–296

Yi D, Zhong X et al (1993) Tertiary of Chinese oil provinces. Petroleum Industry Press, Beijing

Zhao W et al (1997) Comprehensive investigation in petroleum geology. Petroleum Industry Press, Beijing

Zhong X, Zhao C et al (2003) Jurassic in Northern China. Petroleum Industry Press, Beijing

© Petroleum Industry Press and Springer Science+Business Media Singapore 2017

D. Guan et al., *Theory and Practice of Hydrocarbon Generation within Space-Limited Source Rocks*, Springer Geology, DOI 10.1007/978-981-10-2407-8

Printed in the United States
By Bookmasters